The Assessment of Structural Safety

Series in Cement and Concrete Research

Andrew Short, J.P., M.Sc., F.I.C.E.,
F.I.Instruct.E.,C.Eng.,
General Editor

The Assessment of Structural Safety

Ezio Leporati, Ph.D.
Associate Professor of Civil Engineering
Istituto di Scienza delle Costruzioni
Facolta di Ingegneria-Politecnico di Torino

Forthcoming:

Structural Investigations of Ancient Monuments
Theodossius P. Tassios

Structural Aspects of Roman Buildings
Erlio Giangreco

Series in Cement and Concrete Research
Volume I
Andrew Short, J.P., M.Sc., F.I.C.E.,
F.I. Instruct. E., C.Eng.,
General Editor

The Assessment of Structural Safety

A Comparative Statistical Study of the Evolution and Use of Level 3, Level 2, and Level 1

Ezio Leporati

Associate Professor of Civil Engineering
Instituto di Scienza delle Costruzioni
Facolta di Ingegneria-Politecnico di Torino

Translated from original manuscript by Nicoletta Grimoldi.

 RESEARCH STUDIES PRESS

P.O. Box 92
Forest Grove, Oregon 97116

8 Willian Way
Letchworth, Herts
England

The Assessment of Structural Safety
Ezio Leporati, Ph.D.

© Copyright 1979 by Research Studies Press
P.O. Box 92, Forest Grove, Oregon 97116

ISBN: 0-89355-013-2
Library of Congress catalog card number: 79-50708

Printed in the United States of America.

The text was submitted to the Editor for review in July 1977.

Foreword

The use of probabilistic principles and methods in the design and checking of structures, both in the field of civil engineering at large, and civil engineering in particular, had only a very limited number of supporters twenty years ago. Today, it is not only the subject of an ever-wider range of research projects, but it has also been adopted, or is about to be introduced, into the technical codes of nearly all the technologically advanced countries. Structural safety, that is to say, the discipline which deals with these problems, albeit a comparatively young science, has therefore already proved its worth. Its evolution has been similar to that of other disciplines: its general theoretical principles, formulated so as to generate a systematic and rigorous treatment of all structural problems, have soon come up against considerable practical difficulties in routine applications. Thus, the incompleteness of available statistical data does not always allow the rigorous characterization of all the basic random variables involved, stochastic processes turn out to be hard to handle at the application level, and the rigorous assessment of reliability involves serious analytical difficulties. These shortcomings have led to the definition, within the framework of structural safety, of methods characterized by different working levels, depending on the degree of sophistication with which we intend to treat the various aspects of the problem being considered.

The aim of this book is to provide an analytical presentation of the principles and methods that characterize the three design and safety checking levels (referred to as level 3, level 2, and level 1) whose essential features are at present best defined, even though, on account of the constant evolution taking place in this field, they might be still susceptible to refinements.

The book consists of five sections. The first is introductory, providing an overall outline of the issue of safety checking and related problems. The second illustrates the fundamental concepts of the full probabilistic method (level 3). The next section deals with level 2 methods, their theoretical bases and implementation. The fourth section is about the basic concepts of present day semi-probabilistic methods (level 1) and advances a number of proposals for their extension and improvement.

The book is an exhaustive, up-to-date report on structural safety, giving a wealth of detailed bibliographical reference data, viewed in comparison to one another. It is addressed to those who work as structural engineers and possess some basic knowledge of statistical concepts. So as to make the material somewhat more comprehensible to lay readers, in Chapter 5, the author has provided background information on the fundamental definitions, results and working methods of statistics and probability calculation.

We believe that the work of elaboration and synthesis accomplished by Prof. Leporati on the basis of original criteria will foster the dialogue between researchers and designers, helping them to progressively bridge the wide gap between the requirements of practical application and the data that can be gathered from theoretical developments.

Prof. Ing. Franco Levi
h.c. President F.I.P.–C.E.B.

Contents

Introduction

1.1 EVOLUTION OF STRUCTURAL SAFETY CONCEPTS

The evolution of structural safety can be followed in the light of the objectives of design which have been proposed in the course of time. In a qualitative and very "broad" sense, the purpose of design can be conceived to be the attainment of equilibrium between safety and economy. For many years, safety calculations were simply formulated on a deterministic basis according to allowable stress criteria: structures were regarded as safe provided that, within an elastic-linear model, appropriate combinations of the stresses induced at any point by the nominal values of the applied loads did not exceed predefined proportions of the material strengths which were also viewed as deterministic.

The application of this method revealed various shortcomings; attempts to improve it called attention first of all to the need for taking into account the random character of the most important design parameters: namely of the actions, of the strength of the materials, and, at times, of the geometric dimensions. This led to the introduction of probabilistic concepts and operational methods. The first proposals in this direction were advanced by Forsell [1] and Mayer [2] in the first quarter of our century. In the years that followed, other researchers were faced with the same basic problem in their work: how to calculate the probability of failure in a structural element, expressing the failure criterion in these two analogous forms:

$$R - S < 0 \qquad \text{and} \qquad R/S < 1,$$

R being the resistance and S the action, considered as random variables.

Thus, they concluded that sufficiently safe structures should be characterised by a relatively low probability of failure. A suitable criterion for selecting rational values for the failure probability was identified by introducing the concept of "generalized cost", conceived as the initial cost of the structure to which is added the product of the probability of failure and the direct and indirect losses that the occurrence of such failure would entail. In order to optimise a design, the expected generalised cost should be minimized [3,4]. This formulation is attributed to Forsell and Gibrat [5].

A further significant step forward consisted of taking into account the concept of "limit state" [6] conceived as that state beyond which a structure, or part of it, can no longer fulfil the functions or satisfy the conditions for which it was designed. In this manner, all the possible states of behaviour of a structure are taken into consideration; this criterion may be regarded as a generalisation of the concept of failure.

Most modern codes adopt the limit states operational method and divide those into two main groups:
—ultimate limit states corresponding to the maximum load-carrying capacity,
—service limit states related to the criteria governing normal use and durability.

For any predefined limit state, an appropriate design model permits the definition of a safety domain D_X and of its boundary in the space of the random variables $X_1, X_2, \ldots X_n$ (grouped in vector $\underline{X}$) which govern the structural problem.

The reliability of the system, meant as a complement to 1 of the probability p_f of attaining a limit state, is given by the probability that any value of the vector $\underline{X}$ lies within the boundary of the safety domain. As a result:

$$1 - p_f = P\left\{\underline{X} \in D_{\underline{X}}\right\} = \int_{D_{\underline{X}}} f_{\underline{X}}(\underline{x}) \, d\underline{x} \qquad (1.1)$$

where $f_{\underline{X}}(\underline{x})$ denotes the joint probability density of the random variables involved.

For safety checking problems, all those parameters which define the safety domain and the joint probability density of the random variables must be known: the objective is the calculation of reliability.

For design problems, with a given failure probability p_f, at least one of the parameters involved is unknown, and it should be determined by solving (1.1).

The most recently proposed objective of design has been the maximization of the expected utility, and correct theoretical formulations already exist for this [7]. Utility may be defined, in a broad sense, as a unified measure of economic, ethical and social values [8]. In the following, it is assumed to be a known function of the random variables involved.

If we denote the space in which vector $\underline{X}$ is defined by R_n, the utility if $\underline{X} = \underline{x}$ by $V_{\underline{X}}(\underline{x})$, and the joint cumulative distribution function of the random variables by $F_{\underline{X}}(\underline{x})$, then the expected utility $E[V]$ is given as:

$$E[V] = \int_{R_n} V_{\underline{X}}(\underline{x}) \, dF_{\underline{X}}(\underline{x}) \qquad (1.2)$$

In design problems, $F_{\underline{X}}(\cdot)$ and $V_{\underline{X}}(\cdot)$, and therefore $E[V]$, depend upon a given set of decision variables $\underline{\vartheta} = [\vartheta_1, \vartheta_2, \ldots, \vartheta_d]$.

The design objective then consists of the determination of the $\underline{\vartheta}^*$ values, for which:

$$E[V/\underline{\vartheta}^*] = \max_{\underline{\vartheta}} E[V/\underline{\vartheta}] \qquad (1.3)$$

Regions generally exist in the space R_n where $V_{\underline{X}}(\underline{x})$ is positive, regions where it rapidly decreases, and others where it becomes negative. In addition, the utility function is often undulating so that several regions of R_n are marked out which can be idealized as safety domains related to different limit states, each being associated with a different type of structural behavior (e.g., cracking, deflection, failure). The concept of limit state may then be viewed as a discretization of the concept of utility.

Furthermore, for binary utility functions of the type:

$$V_{\underline{X}}(\underline{x}) = \begin{cases} 1 & \text{if } \underline{x} \in D_{\underline{X}} \\ 0 & \text{elsewhere} \end{cases} \qquad (1.4)$$

D_X being a suitable domain in R_n, the calculation of the expected utility is transformed into a reliability calculation in so far as (1.2) coincides with (1.1).

It is likely that in the future it will be possible to formulate in terms of utility solutions to isolated problems of special technical or economic importance. At the present state of knowledge, however, designs and hence codes of practice must aim at more modest objectives, such as a given reliability within standardised operational procedures (for instance, by transforming any given distribution of the random variables involved into normal distributions) or given families of safety indices, since, as W.E.Milne put it, the engineer or the mathematician often "knows how to solve a problem, but he can't do it". However, the choice of optimal values to be prescribed for the probabilities of attaining a given limit state or for reliability indices can be guided by the concept of maximising the expected utility [9] or minimising the generalized cost [3].

Finally it should be noted that the overall reliability of a structure is often affected by events that cannot be treated probabilistically on the basis of current theoretical methods, such as, for instance, gross negligence directly related to lack of professional competence and efficiency of control. Therefore, the greater the overall reliability prescribed by a given code for a class of structures, the lower the prescribed theoretical probabilities of failure must be, and the more rigorous the required standards of professional qualification, of design controls, of acceptance of the materials, and of site inspections must be.

1.2 THE LIMIT STATES METHOD

In dealing with safety checking and design problems in terms of reliability, a limited number of ultimate and service limit states is usually selected for a detailed description of structural behaviour Thus, ultimate limit states can be reached as a result of:
—loss of equilibrium of a part or of the whole of the structure considered as a rigid body,
—failure or excessive plasticity of critical sections of the structure owing to static actions,
—transformation of the structure into a mechanism,
—buckling due to elastic or plastic instability,
—fatigue.
   Service limit states can be reached as a result of:
—excessive deflections,

—excessive vibrations,
—local damages.

Other appropriate limit states can be determined as required, especially if the structure is intended for particular or for unusual functions.

A suitable calculation model, specific for each limit state, allows the definition of a safety domain and of its boundary in the space of the random variables involved. The model should take into account the uncertainties in the basic variables (actions, strengths of materials, geometric dimensions) and in the ensuing response of the structure. The definition of a safety region allows the reliability calculations to be made. The aim of the design established by suitable codes is to meet given levels of safety for certain classes of structures which are intended to be built in accordance with agreed design and construction methods and with suitable means and methods of control.

The choice of safety levels must take into consideration the possible consequences of attaining a given limit state, expressed in terms of:
—risk of loss of human lives,
—the number of lives affected,
—economic consequences.

Thus, as far as ultimate limit states are concerned, the effects of attaining these can be broadly divided into three groups [10]:
—not serious - involving negligible risk of a loss of human lives and modest economic losses,
—serious - involving a moderate risk of a loss of human lives and/or considerable economic losses,
—very serious - involving a high risk of loss of human lives and/or very grave economic losses.

The values of the probabilities of failure usually envisaged for ultimate limit states, with reference to a 50 years design life of the structure, vary between 10^{-3} and 10^{-7}, in relation to the estimated effects listed above.

The probabilities of attaining serviceability limit states, with reference to the same life span as above, are usually chosen in the range 10^{-1} to 10^{-2}.

In addition, to obtain optimal structures under different operational conditions, it is possible to introduce a suitable number of different control classes. A recent proposal [10] defines three different operational classes (I, II, III), corresponding to low, normal and high degrees of control, respectively: these relate to design calculations, production technologies and quality of the materials, and to the structures themselves both during construction and in use. Structures which possess a high degree of reliability should be regarded as belonging to control class III. In this way it is possible not only to reduce the scatter of some of the random variables involved, such as material strengths and geometric dimensions, but also to achieve a marked reduction in the probability of gross errors occurring during the various phases of design, construction, and use of a structure.

Moreover, reasonable probabilities (capable of being expressed statistically as conditional probabilities) should be present to ensure that the structure will not attain ultimate limit states due to the effects of disturbing random events of a discrete type, such as accidental actions (explosions, impacts, acts of war, etc.) and those due to fire. A given structure cannot be expected to fulfil its functions fully if such actions due to extreme causes occur. Damages must be limited, however, by the introduction of appropriate hyperstatic features (e.g., redundant members) in the structural system and by the adoption of suitable protective measures.

1.2.1 The Levels of Verification of the Safety and of the Design

The methods of safety analysis proposed currently for the attainment of a given limit state can be grouped under three basic "levels" (namely levels 3, 2, 1) depending upon the degree of sophistication applied to the treatment of the various problems. A fourth level, characterised by an explicit optimisation of the design at risk, has also been suggested [9]. There are no rigid boundaries between the various levels; the differences among them are operational rather than conceptual.

Level 3 methods require a complete analysis of the problem and also, according to (1.1), the integration, usually multidimensional of the joint distribution density of the random variables extended over the safety domain (or over the complementary failure domain).

Level 2 methods involve simpler calculation procedures since checks are made at a finite number of points of the safety domain boundary and often at one point only. In this way, the multidimensional integration of level 3, which usually involves considerable operational difficulties, is avoided. Random variables are characterized by their known or postulated distribution functions, defined in terms of relevant parameters such as types, means and variances; joint probabilistic behavior is reflected in their covariance matrix. Reliability is expressed in terms of suitable safety indices or, in an equivalent manner, through "operational" failure probabilities, in contrast to "real" or "frequentist" ones (such as can be deduced on the basis of level 3 methods), to indicate that they are affected by a particular set of spproximations and operational assumptions.

Under certain conditions, such operational failure probabilities coincide with the real ones, but generally it is possible to determine only the upper and lower bounds of the appropriate ranges of probability in which the former lie, depending on the convexity properties of the safety domain.

In level 1 methods, the probabilistic aspect of the problem is taken into account by introducing into the safety analysis suitable "characteristic values" of the random variables, conceived as fractiles of a predefined order of the statistical distributions concerned. These characteristic values are associated with partial safety factors that should be deduced from probabilistic considerations so as to ensure appropriate levels of structural reliability. Once the characteristic values are established, it is possible to derive partial safety factors from level 2 methods. In particular, by expressing these factors as continuous functions of the parameters of the basic variables, level 1 methods may become identical with those of level 2 within a given limit state function. In reality, for practical and operational reasons, the same partial safety factors must be used over a range of different design parameters and even for different classes of structural elements. In this manner, practical level 1 methods appear as a discretization of the continuous functions mentioned above, and the reliability of a design will deviate from the target values which are laid down by a given standard code as the objective to be attained. One of the basic problems is that of choosing the partial safety factors so that such deviations are kept within acceptable limits.

It is easy to foresee that, in the immediate future, most engineers will adopt level 1 operational methods and that these methods will form the basis of code provisions.

Level 2 methods may be used by engineers to solve problems of special technical and economic importance, and also by code committees engaged in drafting and revising standard codes of practice to evaluate the partial safety factors required for level 1 procedures.

Level 3 methods are likely to remain in the field of research and to be used for checking the validity of approximations, idealisations, and simplifications that pertain to the lower operational levels.

1.3 AIM AND CONTENTS OF THE PRESENT MONOGRAPH

The aim of this present study is the analysis of the principles and operational procedures which characterise the three levels of design and safety checking mentioned above, in an attempt to give an overall view of the subject. This should enable the reader to use detailed information on single specific problems found in the technical literature even though it is often couched in highly specialised language. Ample use was made of these sources of information in the preparation of this study, in an attempt to highlight the most important problems. An analysis of safety in terms of utility was omitted, and the work was restricted to a study of structural reliability in relation to the limit states method.

Part 2 deals with the basic concepts of the full probabilistic method (level 3). Part 3 presents level 2 operational methods, their theoretical basis, and their probabilistic framework. In Part 4, considerations on level 1 methods are presented.

An analytical treatment of stochastic processes was not included. In view of the difficulty of practical application, this topic is, at present, almost exclusively confined to the field of research. However, in Part 3, the problem of the variability of the actions with time is dealt with within approximations of the same order as those of level 2.

To assist structural designers with a basic, though not necessarily specialised, knowledge of statistical concepts, some statistical problems are described in Part 5, especially with regard to multidimensional distributions and transformations of random variables. The author would be grateful for readers' comments and reflections.

The Probabilistic Limit States Method—Level 3

2.1 PREMISE

Level 3 is the most complete method of reliability analysis. All the random variables involved are represented by their distribution functions; joint probabilisitic behavior is expressed in terms of the covariance matrix. The operational procedure requires knowledge of the appropriate safety domain for each limit state. Taking into account that domain was first proposed in the field of load multipliers [11]; the earliest proposal to use it in structural reliability theory, including geometric and mechanical parameters in addition to actions, was, it seems, suggested by Bolotin [12]. Thanks to the introduction of the concept of the safety domain, we can divide reliability analysis into two main stages: the first stage consists of a deterministic analysis of the behavior of the structure which allows the definition of the safety domain; the second stage consists of numerical ntegration of the joint probability density of the random variables over that domain.

Level 3 methods have been studied by several authors [13, 14, 15, 16, 17, . . .]. These methods, of primary importance in the field of research, are not easily applicable to practical safety checking and design because of the analytical and numerical difficulties involved. They remain, however, the only available means to check the approximations and also the validity of the assumptions which characterise the operational methods of levels 2 and 1.

2.2 THE SAFETY DOMAIN

According to the limit states method, a structure is regarded as unfit for its prescribed purpose when it attains a particular state beyond which it no longer performs its functions or no longer satisfies those conditions it was intended to fulfill. For each limit state, a suitable calculation model can be evolved, founded on the operational methods of the elastic, plastic, or, in a broad sense, nonlinear theory as the case may be. From that model, in the space of parameters of a random character, the safety domain (corresponding to not attaining the limit state) and the complementary failure domain (corresponding to the attainment of the limit state) can be defined, separated by the limit state surface.

Let $\underline{X} = [X_1, X_2, \ldots, X_n]$ denote the vector of basic random variables (strengths of materials, actions, geometric dimensions) involved in the structural problem under study. In a broad sense, with the exception of physical and mathematical constants, all other parameters that are functionally independent may be regarded as basic variables. In the space R_n let us define the limit state function:

$$z = g(x_1, x_2, \ldots, x_n) \qquad (2.1)$$

By hypothesis, z will attain positive values for values of $\underline{X}$ belonging to the safety domain D_n and negative values within the failure domain D'_n.

The limit state surface, representing the boundary of the safety domain, will be expressed analytically by the equation:

$$z = g(x_1, x_2, \ldots, x_n) = 0 \qquad (2.2)$$

All the characteristic features of the structures under consideration are thus contained in and expressed by the domain D_n, which is in turn defined within the range of those parameters which are relevant for safety.

With the probabilistic approach, structures should be designed so as to achieve an optimal reliability level with respect to any limit state. As stated in Section 1.2, that level may vary from one structure to another and it must be established as a function of the risk of loss of human lives, the average number of lives potentially endangered, and the economic consequences, that is, broadly speaking, the general damage to the community. In the following discussion, the reference probability p_f is assumed to be determined in advance and, hence, known for every limit state and for every structure.

2.3 THE CALCULATION OF PROBABILITY OF FAILURE

Let $f_{\underline{X}}(x_1, x_2, \ldots, x_n)$ be the joint probability density function for the $\bar{n}$ random variables involved. Thus:

$$f_{\underline{X}}(x_1, x_2, \ldots, x_n)dx_1 dx_2 \ldots dx_n = P \begin{Bmatrix} x_1 < X_1 \leqslant x_1 + dx_1 \\ x_2 < X_2 \leqslant x_2 + dx_2 \\ \ldots \ldots \ldots \ldots \ldots \\ x_n < X_n \leqslant x_n + dx_n \end{Bmatrix} (2.3)$$

Once this function and the failure domain D'_n are known, the probability p_f of attaining the limit state can be calculated immediately, since this quantity is given by the probability that the vector $\underline{X}$ lies within D'_n. In analytic terms:

$$p_f = \int_{D'_n} f_{\underline{X}}(x_1, x_2, \ldots, x_n)dx_1 dx_2 \ldots dx_n \qquad (2.4)$$

If the n random variables $X_1, X_2, \ldots, X_n$ are, for instance, normally distributed, and have the following mean values:

$$\bar{x}_1, \bar{x}_2, \ldots, \bar{x}_n$$

and the covariance matrix K:

$$K = \begin{bmatrix} \sigma_{11} & \sigma_{12} & \cdots & \sigma_{1n} \\ \sigma_{21} & \sigma_{22} & \cdots & \sigma_{2n} \\ \hline & & \cdots & \\ \sigma_{n1} & \sigma_{n2} & \cdots & \sigma_{nn} \end{bmatrix} \qquad (2.5)$$

in which:

$$\sigma_{ii} = \sigma_i{}^2 = \int_{-\infty}^{+\infty} (x_i - \bar{x}_i)^2 \, f_{X_i}(x_i) \, dx_i \qquad (2.6)$$

are the variances of the marginal distributions $f_{X_i}(x_i) =$

$$\frac{1}{\sqrt{2\pi}\,\sigma_i} \, \exp\left[-\frac{1}{2}\left(\frac{x_i - \bar{x}_i}{\sigma_i}\right)^2\right] \qquad (2.7)$$

and:

$$\sigma_{ij(i \neq j)} = \int_{-\infty}^{+\infty}\int_{-\infty}^{+\infty} (x_i - \bar{x}_i)(x_j - \bar{x}_j)f_{X_i X_j}(x_i, x_j)dx_i dx_j \qquad (2.8)$$

are the covariances, then the joint probability density function will be:
$$f_{\underline{X}}(x_1, x_2, \ldots, x_n) =$$

$$\frac{1}{\sqrt{(2\pi)^n |K|}} \exp\left[-\frac{1}{2} \sum_{i=1}^{n} \sum_{j=1}^{n} \left\{K^{-1}\right\}_{ij} (x_i - \bar{x}_i)(x_j - \bar{x}_j) \right] \qquad (2.9)$$

where $|K|$ denotes the determinant of the covariance matrix (2.5), and $\left\{K^{-1}\right\}_{ij}$ an element of the matrix K^{-1}, inverse value of the matrix K.

In the absence of a correlation between the random variables $X_1, X_2, \ldots, X_n$, the matrices K and K^{-1} become diagonal:

$$|K| = \sigma_1^2 \cdot \sigma_2^2 \cdot \ldots \cdot \sigma_n^2 \qquad (2.10)$$

$$\left\{K^{-1}\right\}_{ij} = \begin{cases} 1/\sigma_i^2 & \text{for } i = j \\ 0 & \text{for } i \neq j \end{cases} \qquad (2.11)$$

and (2.9) is transformed into:

$$f_{\underline{X}}(x_1, x_2, \ldots, x_n) =$$

$$\frac{1}{\sqrt{(2\pi)^n} \, \sigma_1 \sigma_2 \ldots \sigma_n} \exp\left[-\frac{1}{2} \sum_{i=1}^{n} \left(\frac{x_i - \bar{x}_i}{\sigma_i} \right)^2 \right] \qquad (2.12)$$

In other words and with reference to any distribution, if the n random variables are stochastically independent, then the multiplication theorem of probability calculation allows us to express the joint probability density function as the product of the individual densities:

$$f_{\underline{X}}(x_1, x_2, \ldots, x_n) = f_{X_1}(x_1) \cdot f_{X_2}(x_2) \cdot \ldots \cdot f_{X_n}(x_n) . \qquad (2.13)$$

If the variables are statistically dependent and if such dependence can be represented by the covariance matrix, then the correlated variables X_i can be replaced by uncorrelated variables Y_i derived from the former by means of the orthogonal transformation:

$$\underline{y} = U^T \underline{x} \qquad (2.14)$$

in which U^T is the transposed matrix of the eigenvectors relative to the eigenvalues of the covariance matrix. These eigenvalues are the variances of the uncorrelated variables Y_i. By doing so, the simpler analytical expression (2.13) of joint probability density of the random variables holds in any case, and from now on we shall refer to this position unless otherwise specified.

The problem of the safety analysis of a structure consists of the following operations:

I. selection of the parameters that must be treated as random variables and determination of their joint probability density function (statistical problem),

II. taking into account, for every limit state, a space characterised by as many dimensions as the random variables involved, and defining, in such space, the safety and failure domains (deterministic problem),

III. evaluation of the probability of attaining the limit state being considered, by solving a multiple integral extended over the failure domain (analytical numerical problem).

Proceeding in this manner it is possible to obtain the exact reliability of the system. Steps I and III, however, involve considerable difficulties from the operational viewpoint. Even if the direct multi-dimensional integration can be replaced by simulation techniques of the Monte Carlo type, owing to its inherent complexity, this method is restricted to the field of research, and it requires the introduction of drastic simplifications, which characterize level 1 and 2 procedures, if it is to be possible to use it for safety analysis in engineering practice. Yet, this procedure still remains the only available method to check the validity of the approximations, idealisations and simplifications adopted for the lower levels of safety analysis.

2.4 REDUCTION OF THE DIMENSIONS OF THE INTEGRA—TION DOMAIN

The independent random variables $X_1, X_2, \ldots, X_n$ can generally be subdivided into two basic categories. The first of these categories could comprise, for instance, the strengths of materials and the second the actions. Geometric dimensions may belong to either group, but, in a majority of cases, their variability may be neglected (see 2.7). In view of the diverse origins and characteristic features of these random quantities, it might be useful to try to separate their variability in the following manner [15].

Let $f_{X_a}(x_1, x_2, \ldots, x_a)$ be the joint probability density function of a from among the n random variables $(a < n; n - a = p)$, that is of the subvector $\underline{X}_a$ of $\underline{X}_n$. Such variables are referred to as "active".

Let us now evaluate probabilistically the response of the structure in terms of the other p variables $X_{a+1}, X_{a+2}, \ldots, X_n$ (referred to as "passive"), grouped in the subvector $\underline{X}_p$ of $\underline{X}_n$.

For any value: $x_{a+1}, x_{a+2}, \ldots, x_n$ of the p variables, the intersection of the safety domain D_n with the hyperplanes:

$$X_{a+1} = x_{a+1} \quad ,$$
$$X_{a+2} = x_{a+2} \quad ,$$
$$\cdots\cdots\cdots\cdots$$
$$X_n = x_n \quad ,$$

is a subdomain in the space R_a (having reduced dimensions in comparison with those of R_n), that will be denoted by $D_a(\underline{x}_p)$.

Let $G(\underline{x}_p)$ be the conditional probability of attaining the limit state, i.e., the failure probability given that the passive variables assume the predetermined values: $x_{a+1}, x_{a+2}, \ldots, x_n$.

Then:

$$G(\underline{x}_p) = \int_{R_a - D_a(\underline{x}_p)} f(x_1, x_2, \ldots, x_a) dx_1 dx_2 \ldots dx_a \qquad (2.15)$$

Function $G(\underline{x}_p)$ takes on values lying necessarily between 0 and 1, and it can be calculated at points for a sufficient number of different values of the subvector $\underline{X}_p$. It takes into account the random character of the active variables only.

Knowing the joint distribution function of the passive variables:

$$f_{\underline{X}_p}(x_{a+1}, x_{a+2}, \ldots, x_n) \quad ,$$

the probability of attaining the limit state p_f can be evaluated by means of an integration extended to the space R_p only. Then:

$$p_f = \int_{R_p} G(\underline{x}_p) f_{\underline{X}_p}(x_{a+1}, x_{a+2}, \ldots, x_n) dx_{a+1}, dx_{a+2} \ldots dx_n \qquad (2.16)$$

For instance, let X_1, X_2, X_3 be three independent random variables with generic distribution functions:

$$f_{X_1}(x_1) \ , \ f_{X_2}(x_2) \ , \ f_{X_3}(x_3) \ ,$$

respectively. The safety domain D_3 will be defined in the space R_3. Considering X_3 as a passive variable and denoting the intersection of the safety domain D_3 with the plane $X_3 = x_3$ by $D_2(x_3)$, (2.15) results in:

$$G(x_3) = \iint_{R_2 - D_2(x_3)} f_{X_1}(x_1) \cdot f_{X_2}(x_2) dx_1 dx_2$$

and (2.16) becomes:

$$p_f = \int_{-\infty}^{+\infty} G(x_3) f_{X_3}(x_3) dx_3 \ .$$

Assuming instead that X_1 and X_2 are passive variables and denoting the intersection of D_3 with the planes:

$$X_1 = x_1 \ ,$$
$$X_2 = x_2 \ ,$$

by $D_1(x_2, x_3)$, then:

$$G(x_1, x_2) = 1 - \int_{D_1(x_1, x_2)} f_{X_3}(x_3) dx_3 \ ,$$

$$p_f = \int_{-\infty}^{+\infty} \int_{-\infty}^{+\infty} G(x_1, x_2) f_{X_1}(x_1) f_{X_2}(x_2) dx_1 dx_2 \ .$$

Taking the $G(\cdot)$ function into account is of special interest if the data pertaining to the random variables are available for different times or situations. Suppose, for example, that a type of structure is produced in some quantity from a given "center" and is then utilized under slightly varying conditions by a number of "branches", that is, with variations of the statistics of p among the n variables. Assuming that the a = n − p variables common to all the structures are active, the resolution of (2.15), which usually involves considerable numerical complexity, will be performed only once at the production center, while the simpler calculations needed for (2.16) will be effected case by case by the branches on the basis of their knowledge of the distributions of the local factors.

2.5 OPERATING SPACES

In safety analysis it is necessary to decide in which space of the random variables the problem has to be treated. It is common practice to denote the actions, the material properties and geometric dimensions as "basic variables" which define the "input space". In most cases, however, safety domains are given in the space of the "output variables", e.g. the effects of the actions and the resistances of cross sections or of the structural elements. In fact, the engineer prefers to operate in this space, since its dimensions are usually reduced with respect to those of the input space.

Furthermore, for some problems, it might be necessary to introduce other variables, such as, for instance, stresses and strains, which define the "state space". The three groups of variables are related to one another by appropriate functional relations, that are generally nonlinear.

The probabilistic analysis of safety within a level 3 framework can be performed in any of these spaces with equal validity.

If we denote the basic variables by $\underline{X}$ and the output variables by $\underline{Y}$, before performing any reliability calculation, the relation $x_h = g_{X_h}(\underline{y})$ must be used to transform the safety domain from output space to input space (analytical problem), in order to check safety in the space of the basic variables.

Operating in the output space in which the safety domain is defined, the functional relations $y_k = g_{Y_k}(\underline{x})$ must be used to calculate the joint probability density function of the output variables (analytical-statistical problem, more complicated than the previous one).

Finally, operating within the state space [37], both the basic variables and the safety domain must be transformed.

2.6 THE LIMIT STATE FUNCTION

The safety domain does not necessarily have to be expressed in explicit form, although this is usually feasible in the most common cases. Examples of implicit definition of the safety domain are encountered in a number of engineering problems, such as: failure criteria involving plastic mechanisms, instability modes, maximum deformations in nonlinear systems, maximum widths of cracks.

Moreover, it may not always be possible to express (2.1) by a single analytical function z over the entire region of definition of the random variables.

However, in several cases commonly relevant to practical applications, the limit state equation (2.2) can be expressed in this simple form:

$$g_R(x_1, x_2, \ldots, x_m) - g_S(x_{m+1}, x_{m+2}, \ldots, x_n) = 0 \qquad (2.17)$$

where:

$g_R(\cdot)$ and $g_S(\cdot)$ denote appropriate functional relations,

$X_1, X_2, \ldots, X_m$ denote the strengths of materials,

$X_{m+1}, X_{m+2}, \ldots, X_n$ denote the actions.

The geometric dimensions may appear in both the functions $g_R(\cdot)$ and $g_S(\cdot)$. Assuming that:

$$r = g_R(x_1, x_2, \ldots, x_m) , \qquad (2.18)$$

and $\qquad s = g_S(x_{m+1}, x_{m+2}, \ldots, x_n) , \qquad (2.19)$

the limit equation takes on, in these cases, the classical form:

$$r - s = 0 , \qquad (2.20)$$

and it is represented by a straight line in the plane of the random variables R and S.

Generally, various equivalent limit state functions $z = g(x_1, x_2, \ldots, x_n)$ may exist, that is various functions corresponding to the same failure domain $D'_n (z < 0)$. Thus, when dealing with problems involving only two random variables R and S, besides the classical form:

$$z = r - s , \qquad (2.21)$$

the limit state function can also be expressed by the equivalent form:

$$z = (r - s)^3 \;\; ; \;\; z = (r - s)^5 \;\; ; \text{etc.}$$

and, for positive random variables R and S, also by the usual reference functions:

$$z = (r/s) - 1 \;\; ; \;\; z = \ln (r/s) ,$$

(see Fig. 3.1).

Operating with level 3 methods, any equivalent limit state may be adopted, since the same failure domain being associated with

each of these, the value of p_f calculated on the basis of (2.4) will
not change.

2.7 STATISTICAL DATA

The validity of probabilistic methods of safety analysis depends
to a considerable extent upon the realism of the statistical assumptions relating to the variables involved. The definition of the probability distributions of the basic variables is based on the statistical
interpretation of the available data and on considerations both
physical (concerning, for instance, upper and/or lower bounds)
and mathematical (for instance, extreme character). Thus theoretical considerations, previous experience with fitting the strengths
of similar materials, and the results of local or international inquiries
on actions usually allow the designer to know "a priori" the type
of distribution function that must be selected for the random variable under consideration. The parameters of these distributions
must usually be determined, even in such cases, through a statistical elaboration of the specific experimental data.

Where, as is often the case, the number of available observations
is limited, a use of "Bayesian" estimating procedures [18,19] to
evaluate such statistical parameters is of considerable interest.

With regard to the types of distribution functions of the basic
variables, the following considerations may serve as a general
guideline.

The strengths of most ductile materials are usually well represented by normal and log-normal distributions. The latter distribution function should be preferred to the gaussian, since it also
satisfies the practical requirement of being defined in the field of
positive argumental values only. For mechanical properties of brittle materials, the data fit the Weibull distributions [20] well.

Normal and log-normal distributions are also generally selected
to represent the variability of geometric dimensions. Having modest
scatter in comparison to that of the strengths of materials and of
the actions, these values can be treated usually on a deterministic
basis, except when dealing with elements that are particularly sensitive to geometric imperfections, such as slender beams in compression or thin slabs.

A considerable contribution to the probabilistic definition of
actions and to their representation in an idealized form was made
by J.C.S.S. CEB - CECM - CIB - FIP - IABSE - RILEM [21], which

initiated specific inquiries and produced an excellent collection of statistical information on permanent loads and variable actions, presented within the framework of a unified methodology. Broadly, it can be said that the densities of building materials and a considerable part of the imposed deformations can be represented by normal distributions. Variable actions fit extreme distributions of type I; seismic actions are an exception; for these an extreme distribution of type II is more appropriate.

For the determination of statistical distributions of quantities depending upon one or more basic variables, the probabilistic theory of transformation of the variables (see 5.5) or simulation techniques of the Monte Carlo type (see 5.7) can be used.

2.8 THE FUNDAMENTAL PROBLEM

Let the limit state function (2.1) be of the type:

$$z = r - s \qquad (2.21)$$

In it, let R and S be two random variables expressed in the same units of measurement and such as to take on argumental values of the same sign, for instance, positive.

The limit state straight line, from equation:

$$z = r - s = 0 \qquad (2.20)$$

is the bisecting line of the axes r and s. The failure domain D', the shaded area in Fig. 2.1, is the portion of the plane enclosed between that defined by (2.20) and the positive half of the axis s. The probability p_f of attaining the limit state is given by the probability that the determination (r,s) belongs to the failure domain D', i.e., that the random variable Z takes on negative values.

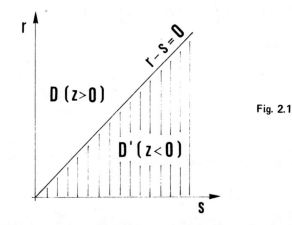

Fig. 2.1

If we denote the joint probability density function of the variables under study by $f_{RS}(r,s)$, then it follows from (2.4):

$$p_f = \iint_{D'} f_{RS}(r,s) dr \, ds \ . \tag{2.22}$$

Assuming R and S to be stochastically independent, (2.22) becomes:

$$p_f = \iint_{D'} f_R(r) \cdot f_S(s) dr \, ds \ , \tag{2.23}$$

where $f_R(r)$ and $f_S(s)$ denote the probability density of R and S respectively.

The domain D' is a normal region with respect to the lines parallel to the r and to the s axis.

Integrating along the vertical strips, we obtain:

$$p_f = \int_0^\infty f_S(s) \left[\int_0^s f_R(r) dr \right] ds = \int_0^\infty F_R(s) f_S(s) ds \ , \tag{2.24}$$

where:

$$F_R(s) = \int_0^s f_R(r) dr \tag{2.25}$$

is the cumulative distribution function of the random variable R.

Integrating along horizontal strips, the equivalent expression is:

$$p_f = \int_0^\infty f_R(r) \left[\int_r^\infty f_S(s) ds \right] dr = \int_0^\infty f_R(r) \left[1 - F_S(r) \right] dr \ , \tag{2.26}$$

where:

$$F_S(r) = \int_0^r f_S(s) \, ds = 1 - \int_r^\infty f_S(s) \, ds \tag{2.27}$$

is the cumulative distribution function of S.

Expressions (2.24) and (2.26) represent the classical relationships for the solution of one-dimensional problems of structural safety (convolution formulae) and they can also be easily deduced by means of elementary considerations of probability calculation [3, 22].

Once the statistical distributions of the random variables R and S are known, the solution, usually numerical, of one of the two integrals (2.24) or (2.26) permits the determination of p_f.

When the random variables R and S, though having the same

sign simultaneously, can attain either positive or negative argumental values, the failure domain D' is extended to the entire portion of plane comprised between the bisecting line $r - s = 0$ and the axis s (see Fig. 2.2). In these cases, while (2.22) and (2.23) retain their validity with this actual meaning of D', it is necessary to add the contribution represented by the integration extended as well to negative argumental values to (2.24) and (2.26). In addition, $f_R(r)$ generally assumes different analytical forms in the ranges $-\infty < r < 0$ and $0 < r < \infty$.

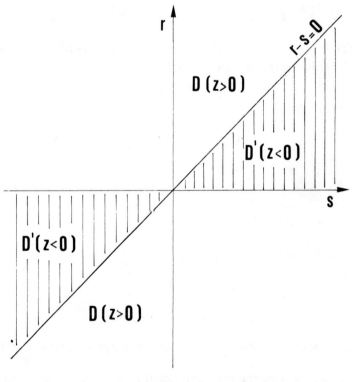

Fig. 2.2

2.8.1 Normal Distributions of R and S

Of special importance is the case where the two random variables R and S are both gaussian, because this type of distribution is commonly used in practice. With such assumptions, the calculation of p_f is relatively easy.

Let: $\bar{r}, \bar{s}, \sigma_R, \sigma_S$ be the mean values and the standard deviations of the independent variables R and S.

The random variable Z defined in (2.21) will also be normal, with parameters:

$$\bar{z} = \bar{r} - \bar{s} \; ; \; \sigma_z = \sqrt{\sigma_R^2 + \sigma_S^2} \; .$$

The probability of failure, p_f, may be expressed in the form:

$$p_f = P\left\{Z<0\right\} = F_Z(0) = \frac{1}{\sigma_z\sqrt{2\pi}} \int_{-\infty}^{0} \exp\left[-\frac{(z-\bar{z})^2}{2\sigma_z^2}\right] dz \; , (2.28)$$

where $F_Z(\cdot)$ denotes the cumulative normal distribution function.

Introducing the standardized variable:

$$u = (z - \bar{z})/\sigma_Z \; ,$$

(2.28) becomes:

$$p_f = \frac{1}{\sqrt{2\pi}} \int_{-\infty}^{-\bar{z}/\sigma_Z} e^{-u^2/2} \, du =$$

$$\frac{1}{\sqrt{2\pi}} \int_{-\infty}^{-\beta} e^{-u^2/2} \, du = \Phi(-\beta) = 1 - \Phi(\beta) \; , \qquad (2.29)$$

having assumed that:

$$\beta = \frac{\bar{z}}{\sigma_z} = \frac{\bar{r} - \bar{s}}{\sqrt{\sigma_R^2 + \sigma_S^2}} \qquad (2.30)$$

and denoting the standardized cumulative normal distribution by $\Phi(\cdot)$.

Thus, knowing the parameters of the distributions of R and S, the β coefficient is calculated by means of (2.30) and, referring to (2.29), the value of p_f can be read on the tables of the standardised normal cumulative distribution function. Some of the values of the relationship between p_f and β in (2.29) are listed in Table 2.1.

Table 2.1

β	$p_f = \Phi\,(-\beta)$	Limit states involved
1.282	10^{-1}	service
2.326	10^{-2}	
3.090	10^{-3}	ultimate
3.719	10^{-4}	
4.265	10^{-5}	
4.753	10^{-6}	
5.199	10^{-7}	

2.8.1.1 Geometric interpretation of the coefficient β

The *reliability index* β defined in (2.30) can be interpreted geo-metrically as follows:

Standardising the random variables R and S, by stating:

$$\xi = (r - \bar{r})/\sigma_R \ ,$$
$$\eta = (s - \bar{s})/\sigma_S \ ,$$

(2.20) becomes:

$$z = \xi\sigma_R - \eta\sigma_S + \bar{r} - \bar{s} = 0 \qquad (2.31)$$

The distance 0P of the straight line (2.31) from the origin 0 of the axes ξ,η (Fig. 2.3) is given by:

$$0P = \frac{\bar{r} - \bar{s}}{\sqrt{\sigma_R{}^2 + \sigma_S{}^2}}$$

Taking (2.30) for any distribution of R and of S into account, one gets: $0P = \beta$.

For the case where R and S are both normal, equation (2.29) gives:

$$0P = \beta = -\ \Phi^{-1}(p_f) \qquad (2.32)$$

where $\Phi^{-1}(\cdot)$ denotes the inverse function of the standardised cumulative normal distribution.

Thus, in the plane of the standardised normal variables $\xi,\ \eta$, the coefficient β represents the distance of the limit state straight line from the origin; from equation (2.32) this distance becomes, in turn, a measure of design reliability.

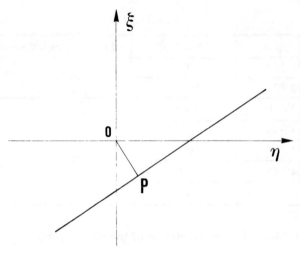

Fig. 2.3

2.8.1.2 The central safety factor

The central safety factor:

$$\gamma_0 = \bar{r}/\bar{s} \tag{2.33}$$

is the ratio between the mean values of R and S.

With reference to (2.29) we can write:

$$\beta = -\Phi^{-1}(p_f) = \frac{\bar{r} - \bar{s}}{\sqrt{\sigma_R^2 + \sigma_S^2}} = \frac{(\bar{r}/\bar{s}) - 1}{\sqrt{\dfrac{\sigma_R^2}{r^{-2}}\dfrac{r^{-2}}{s^{-2}} + \dfrac{\sigma_S^2}{s^{-2}}}} \tag{2.34}$$

Denoting the coefficients of variation of R and S by:

$$\begin{aligned} c_R &= \sigma_R/\bar{r} \\ c_S &= \sigma_S/\bar{s} \end{aligned} \tag{2.35}$$

(2.34) becomes:

$$\beta = -\Phi^{-1}(p_f) = \frac{\gamma_0 - 1}{\sqrt{c_R^2 \gamma_0^2 + c_S^2}} \tag{2.36}$$

After squaring and rearranging equation (2.36):

$$\gamma_0 = \frac{1 + \sqrt{\beta^2(c_R^2 + c_S^2) - \beta^4 c_R^2 c_S^2}}{1 - \beta^2 c_R^2} \tag{2.37}$$

Equation (2.37) expresses the functional connection between γ_0 and β and, therefore, between γ_0 and p_f; the corresponding diagrams were plotted by several authors (for instance [22]) for a certain number of values of c_R and c_S. An examination of equation (2.37) leads to a realisation of the importance that must be attached with reference to safety to a high value of the coefficient of variation of R, an ascending function of the scatter of the strength of materials.

Actually, assuming that R and S are normal, when c_R is equal to or greater than $1/\beta$, it is no longer possible to obtain the reliability level associated with that chosen value of β. In fact, as indicated in Section 2.7, the assumption that both random variables R and S are normal is not justified in a broad sense. The situation outlined above can be attenuated by assuming more realistic probability distributions (see Fig. 2.5).

2.8.2 Log-Normal Distributions of R and S

A random variable X is defined as log-normal when its logarithms ln x are normally distributed. The mechanical properties of ductile materials are in general well fitted by such a distribution which is characterised by only positive argumental values. These considerations justify its wide use for structural safety calculations, at least to express the distribution of the resistances. If X has log-normal distribution, then:

$$\sigma_{\ln x}^2 = \ln (c_X^2 + 1) \qquad (2.38)$$

$$\overline{\ln x} = \ln \bar{x} - \frac{1}{2} \ln (c_X^2 + 1) \qquad (2.39)$$

Writing the limit state function in the form:

$$z = \ln r - \ln s , \qquad (2.40)$$

if R and S are log-normal, then the variable Z is normal.

Assuming again that $\beta = \bar{z}/\sigma_z$ and operating as in Section 2.8.1, the failure probability p_f is:

$$p_f = P \left\{ Z < 0 \right\} = F_Z (0) = \Phi (-\beta) .$$

Since:

$$\bar{z} = \ln \left[\frac{\bar{r}}{\bar{s}} \Big/ \left(\frac{c_R^2 + 1}{c_S^2 + 1} \right)^{1/2} \right] ,$$

$$\sigma_Z = [\ln(c_R^2 + 1) + \ln(c_S^2 + 1)]^{1/2} ,$$

then:

$$\beta = \frac{\ln[\gamma_0 / \left(\dfrac{c_R^2 + 1}{c_S^2 + 1}\right)^{1/2}]}{[\ln(c_R^2 + 1)(c_S^2 + 1)]^{1/2}}$$

and hence:

$$\gamma_0 = \sqrt{\frac{c_R^2 + 1}{c_S^2 + 1}} \ \exp{[\beta \sqrt{\ln(c_R^2 + 1)(c_S^2 + 1)}]} \qquad (2.41)$$

Should c_R and c_S be sufficiently small with respect to unity, (2.41) can be written in the approximate form:

$$\gamma_0 = \exp{[\beta \sqrt{c_R^2 + c_S^2}]} . \qquad (2.42)$$

Expression (2.42) is generally called the "exponential form", and it has been preferred to (2.37) by a number of authors and code drafting committees [9, 23, 24, 25], within the operational range of the "first order — second moment" theory (see Part 3).

2.8.3 Any Given Distribution of R and S

For distributions of any type of the random variables R and S, the value of p_f as well as the curves $p_f - \gamma_0$ for predetermined values of c_R and c_S must be obtained numerically on the basis of (2.24) or (2.26).

Fig. 2.4, taken from [26], reproduces a fairly large number of curves $p_f - \gamma_0$ for different types of distributions of R and S: normal, log-normal, and the extreme of type I. This last one is characterized by the following distribution function:

$$F(x) = \exp\left\{ - \exp{[-a(x - u)]} \right\} \qquad (2.43)$$

where:

$$\bar{x} = u + 0.577/a ,$$

$$\sigma_X = 1.2825/a .$$

Fig. 2.5, with reference to the types of distribution functions considered in Fig. 2.4, reproduces some $\gamma_0 - c_R$ curves for $\beta = 4$ and for predetermined values of c_S (0.1 and 0.3). From these, the greater influence of high values of c_R on the failure probability can be shown, where R and S are both normal, in comparison with other types of distribution functions, all other conditions being equal.

Curve	Load		Strength		Symbol
	Distribution	V_S	Distribution	V_R	
1	Normal	0.10	Normal	0.10	————
2		0.30		0.10	
3	Lognormal	0.10	Lognormal	0.10	– – – –
4		0.10		0.30	
5		0.30		0.10	
6		0.30		0.30	
7	Normal truncated	0.10	Lognormal	0.10	–·–·–
8		0.10		0.30	
9		0.30		0.10	
10		0.30		0.30	
11	Extreme type I	0.10	Lognormal	0.10	— – — –
12		0.10		0.30	
13		0.30		0.10	
14		0.30		0.30	

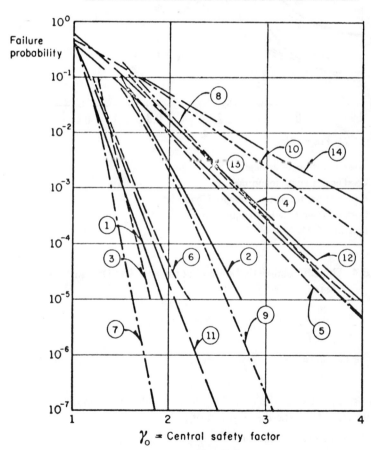

Fig. 2.4 — From [26]

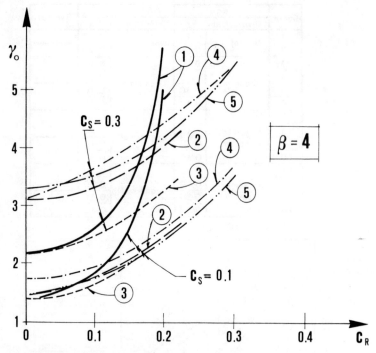

Fig. 2.5 — From [9]

1 Load and strength normal
2 Load and strength lognormal
3 Load normal truncated; strength lognormal
4 Load extreme type I; strength lognormal
5 Equation 2.42

2.8.4 The Characteristic Safety Factor and The Design Safety Factor

In addition to γ_0, other common reference parameters for structural safety are:

—the characteristic safety factor

$$\gamma_k = r_{0.05}/s_{0.95} \quad , \tag{2.44}$$

the ratio between the 0.05 fractile of the distribution of R and the 0.95 fractile of the distribution of S,

—the design safety factor:

$$\gamma^* = r_{0.005}/s_{0.95} \tag{2.45}$$

the ratio between the 0.005 fractile of the distribution of R and the 0.95 fractile of the distribution of S.

When both random variables R and S are normal, then:

$$\gamma_k = \frac{\bar{r}\,(1 - 1.645\,c_R)}{\bar{s}\,(1 + 1.645\,c_S)} = \gamma_0\,\frac{1 - 1.645\,c_R}{1 + 1.645\,c_S} , \tag{2.46}$$

$$\gamma^* = \frac{\bar{r}\,(1 - 2.576\,c_R)}{\bar{s}\,(1 + 1.645\,c_S)} = \gamma_0\,\frac{1 - 2.576\,c_R}{1 + 1.645\,c_S} . \tag{2.47}$$

A number of diagrams showing the curves of γ_k and of γ^* as functions of p_f for predetermined values of c_R and c_S are given in Fig. 2.6 and Fig. 2.7 respectively.

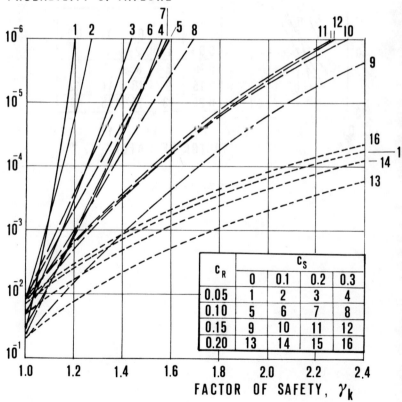

Fig. 2.6 — From [22]

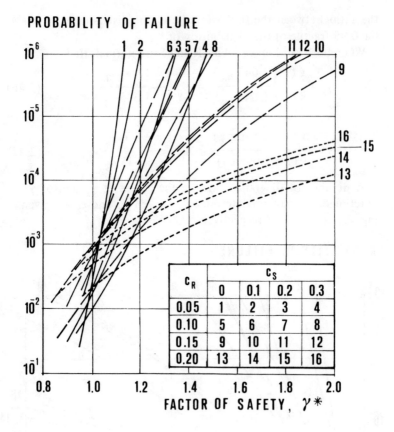

Fig. 2.7 — From [22]

Level 2 Methods

3.1 PREMISE

As shown in Part 2, level 3 requires the use of sophisticated techniques for the creation of accurate statistical models, and the integration, usually multidimensional, of the joint probability density function of the random variables. These operations are certainly not easy to perform in routine design practice, except where the safety domain boundaries and distribution functions of the random variables are analytically simple and the number of the dimensions of the operating space is reduced (not more than three).

These considerations gave rise in time to progressive and continuous research for simpler procedures, with the aim of being able to obtain an approximate measure of the reliability of structures with algebraic operational methods, generally of an iterative character.

With level 2 methods, safety checks are made at a finite number of points (often at one point only) of the safety domain boundary which is projected into the space of noncorrelated and standardised basic random variables, thus avoiding direct multidimensional integration or the use of simulation techniques. The parameters that must be determined are the distances of appropriately selected points of the safety domain boundary from the origin of the system of coordinates mentioned above, in the case of multiple checks, or the minimum distance, in the case of one check only. With the assumption of joint normal distribution for the random variables involved and of suitable geometries for the safety domain boundary (hyperplanes or hyperspheres), it is possible to associate a precise meaning in terms of reliability with this distance; for boundaries of any form, upper and lower bounds can be determined within which the actual reliability lies; these are more or less close depending upon the convexity characteristics of the safety domain.

These considerations alone would justify a preference for being able to deal with normal distributions of the random variables, and a tendency to transform all other types of distribution into such. Furthermore these transformations should be analytically simple and such as to alter the statistical information content involved in the safety analysis of a specific problem as little as possible.

Within the context of level 2 code proposals, operational schemes have been suggested [27, 10] that provide for safety checks only at the point of minimum distance from the origin of the normalised and standardised variables, associating with such distance the "operational" values of reliability which can be justified by approximating the safety domain boundary by a hyperplane which, at the point under consideration, is tangential to the hypersphere having its center at the origin. For these operational values of reliability it is not necessary to insist on a rigorous probabilistic interpretation (i.e., in the sense of relative frequencies of real non-failures), but only on establishing the significance of the compared measures of risk within standardized operational procedures. Several authors have contributed to the elaboration of present-day positions, through step by step modifications of the basic "mean first order—second moment" concepts.

The present Part 3 deals with the evolution of the fundamental concepts of level 2. The subdivision of the subject matter is at times subjective and arbitrary, especially with regard to the inclusion of some procedures in the nonparametric group or in the group which makes use of distributional assumptions: at times the original proposal had been formed in the opposite group.

3.2 NON—PARAMETRIC "SECOND MOMENT" METHODS

The basic concept of second moment methods is to set out all random quantities in terms of their first two moments. These operational procedures were introduced for codification purposes by Cornell within the context of the so-called "mean first order—second moment" method (Section 3.2.1).

A number of authors have made various specific contributions to this operational field; an improved version of the method was proposed by Lind and Hasofer (Section 3.2.2).

3.2.1 The Mean First Order—Second Moment Method (MFOSM)

The philosophical bases of the MFOSM method were already present in the works of Mayer [2], Rzhanitsyn [28], and Basler [29]; it was Cornell who gave them an organic form, suited to practical applications [30], with his formulation of the problem in the space of output variables.

This is a "second moment" type theory, insofar as the statistical character of the uncertain quantities involved in safety analysis is described by their means and variances, that is, by only the first two moments of their perhaps unknown probability distributions. Joint probabilistic behavior is expressed by the correlation coefficients; the moments of higher order are not used. No assumption is made as to the type of distribution of the variables.

The limit state function chosen as reference by Cornell is:

$$z = r - s \ , \tag{3.1}$$

where the variables R and S denote the effects of the material strengths and of the actions respectively, expressed by means of the functional relations:

$$r = g_R (x_1, x_2, \ldots, x_m) \tag{3.2}$$

$$s = g_S (x_{m+1}, x_{m+2}, \ldots, x_n) \tag{3.3}$$

These relations are represented approximately as linear functions by expanding them in a Taylor series corresponding to the mean values $\bar{x}_i$ of the basic variables X_i, neglecting nonlinear terms (mean first order).

Then:

$$r \cong g_R (\bar{x}_1, \bar{x}_2, \ldots, \bar{x}_m) + \sum_{i=1}^{m} a_i (x_i - \bar{x}_i) \tag{3.4}$$

where:

$$a_i = \left[\frac{\partial g_R (x_1, x_2, \ldots, x_m)}{\partial x_i} \right]_{\substack{x_1 = \bar{x}_1 \\ x_2 = \bar{x}_2 \\ \cdots \\ x_m = \bar{x}_m}} \tag{3.5}$$

The mean and variance of R, within the mean first order method are given by:

$$\bar{r} = g_R(\bar{x}_1, \bar{x}_2, \ldots, \bar{x}_m) \qquad (3.6)$$

$$\sigma_R^2 = \sum_{i=1}^{m} \sum_{j=1}^{m} a_i a_j \, \rho_{X_i X_j} \, \sigma_{X_i X_j} \qquad (3.7)$$

where $\rho_{X_i X_j}$ is the correlation coefficient of X_i and X_j.

If the variables $X_1, X_2, \ldots, X_m$ are independent, (3.7) becomes:

$$\sigma_R^2 = \sum_{i=1}^{m} a_i^2 \, \sigma_{X_i}^2 \,. \qquad (3.8)$$

Relations analogous to (3.4) to (3.8) are also valid for the variable S, expressed by the functional relation (3.3).

These results enable us to calculate the mean and variance of Z; where R and S are independent, then:

$$\bar{z} = \bar{r} - \bar{s} \,, \qquad (3.9)$$

$$\sigma_Z^2 = \sigma_R^2 + \sigma_S^2 \,. \qquad (3.10)$$

In the context of level 3 safety checking, as seen in Section 2.8, it is necessary to assess the probability of the failure event, i.e. $P\{Z<0\}$, and to check that this probability is lower than a predetermined p_f value.

With the MFOSM method, the verification:

$$[P\{Z < 0\}] < p_f \qquad (3.11)$$

is replaced by a criterion which only takes into account the mean value and the standard variation of Z; therefore, it is only necessary to check that:

$$\bar{z} - \beta \, \sigma_Z \geqslant 0 \,, \qquad (3.12)$$

That is, reliability is measured by the number of standard deviations by which the mean of Z exceeds 0. That number β is denoted as the "design reliability index" and its values must be established in advance by suitable specifications or code provisions.

On the basis of (3.9) and (3.10), (3.12) becomes:

$$\frac{\bar{r} - \bar{s}}{\sqrt{\sigma_R^2 + \sigma_S^2}} \geqslant \beta \qquad (3.13)$$

and, after introducing the central safety coefficient γ_0, and then squaring and rearranging:

$$\frac{\bar{r}}{\bar{s}} = \gamma_0 \geqslant \frac{1 + \beta\sqrt{c_R^2 + c_S^2 - \beta^2 c_R^2 c_S^2}}{1 - \beta^2 c_R^2} \qquad (3.14)$$

which is a relationship analogous to (2.37) but expressed here in accordance with the MFOSM method.

It is obvious that, with the same theory, it is possible to start out from various formulations of the limit state function. Thus, writing [31]:

$$z = \ln r - \ln s , \qquad (3.15)$$

it is possible to express the relationship for the verification of safety in an exponential form:

$$\frac{\bar{r}}{\bar{s}} = \gamma_0 \geqslant \exp\left(\beta\sqrt{c_R^2 + c_S^2}\right), \qquad (3.16)$$

This form has been adopted in Mexican [31] and Canadian [25] code proposals.

3.2.1.1 Formulation of safety analysis in the space of the basic variables

Working in the space of the basic variables X_i, let

$$z = g(x_1, x_2, \ldots, x_n) \qquad (3.17)$$

be the limit state function, expressed so that the locus of the points for which Z takes on negative values corresponds to the failure domain.

The treatment effected above in the space of the output variables R and S is equivalent to the expansion in power series of (3.17) in correspondence with the mean values of the basic variables, neglecting nonlinear terms.

In that case

$$z' = g(\bar{x}_1, \bar{x}_2, \ldots, x_n) + \sum_{i=1}^{n} \left(\frac{\partial g}{\partial x_i}\right)_{\bar{x}} (x_i - \bar{x}_i) \qquad (3.18)$$

where the notation:

$$\left(\frac{\partial g}{\partial x_i}\right)_{\bar{x}}$$

indicates that the partial derivatives are calculated with the mean values $\bar{x}_1, \bar{x}_2, \ldots, \bar{x}_n$.

If the variables X_i are independent, then:

$$\bar{z}' = g(\bar{x}_1, \bar{x}_2, \ldots, \bar{x}_n)$$

$$\sigma_{Z'} = \sqrt{\sum_{i=1}^{n} \left[\left(\frac{\partial g}{\partial x_i} \right)_{\bar{x}} \sigma_{X_i} \right]^2}$$

and the design equation becomes:

$$\beta = \bar{z}'/\sigma_{Z'} = \frac{g(\bar{x}_1, \bar{x}_2, \ldots, \bar{x}_n)}{\sqrt{\sum_{i=1}^{n} \left[\left(\frac{\partial g}{\partial x_i} \right)_{\bar{x}} \sigma_{X_i} \right]^2}} \qquad (3.19)$$

If we set:

$$a_i = -\left(\frac{\partial g}{\partial x_i} \right)_{\bar{x}} \sigma_{X_i} \bigg/ \sqrt{\sum_{i=1}^{n} \left[\left(\frac{\partial g}{\partial x_i} \right)_{\bar{x}} \sigma_{X_i} \right]^2} \qquad , \quad (3.20)$$

then:

$$\sigma_{Z'} = -\sum_{i=1}^{n} a_i \left(\frac{\partial g}{\partial x_i} \right)_{\bar{x}} \sigma_{X_i} \qquad (3.21)$$

and (3.19) can be written in the form:

$$g(\bar{x}_1, \bar{x}_2, \ldots, \bar{x}_n) + \beta \sum_{i=1}^{n} a_i \left(\frac{\partial g}{\partial x_i} \right)_{\bar{x}} \sigma_{X_i} = 0 \qquad (3.22)$$

If (3.17) is linear, that is of the type:

$$z = g(x_1, x_2, \ldots, x_n) = a_1 x_1 + a_2 x_2 + \ldots + a_n x_n \quad , \qquad (3.23)$$

then:

$$\bar{z} = \sum_{i=1}^{n} a_i \bar{x}_i \quad , \quad \left(\frac{\partial g}{\partial x_i} \right)_{\bar{x}} = a_i \quad ,$$

$$a_i = -a_i \sigma_{X_i} \bigg/ \sqrt{\sum_{i=1}^{n} (a_i \sigma_{X_i})^2} \quad ,$$

$$\sigma_Z = -\sum_{i=1}^{n} a_i a_i \sigma_{X_i} \ .$$

In this case, (3.22) becomes:

$$\sum_{i=1}^{n} a_i(\bar{x}_i + a_i \beta \sigma_{X_i}) = 0 \ . \tag{3.24}$$

With reference to the presence of only two variables R and S and to the limit state function z = r − s, since:

$$a_R = 1 \ , \quad a_S = -1 \ ,$$

$$a_R = -\sigma_R/\sqrt{\sigma_R^2 + \sigma_S^2} \ , \quad a_S = +\sigma_S/\sqrt{\sigma_R^2 + \sigma_S^2} \ ,$$

then, on the basis of (3.24):

$$\bar{r} + a_R \beta \sigma_R - \bar{s} - a_S \beta \sigma_S = 0 \ ,$$

that is:

$$\bar{r}(1 + a_R \beta c_R) - \bar{s}(1 + a_S \beta c_S) = 0 \ , \tag{3.25}$$

expressions equivalent to (3.13).

Within the model R, S, equation (3.25) represents the design relationship written in terms of partial safety factors.

Setting:

$$\Delta_R = 1 + a_R \beta c_R \ ,$$

$$\Delta_S = 1 + a_S \beta c_S \ ,$$

then:

$$\Delta_R \bar{r} - \Delta_S \bar{s} = 0 \ . \tag{3.26}$$

Since the Δ_R and Δ_S factors can be regarded, with acceptable approximation, as being constant within a fairly ample range of the ratio c_R/c_S, the development of level 1 methods based on the use of partial safety factors can be shown to be justified (see Part 4).

3.2.1.2 Calibration of the reliability index β

In the context of the second moment method, no assumption is made about the statistical distribution of the random variables; hence, the distribution of Z is also undefined. Consequently, the information on reliability associated with a given value of $\beta = \bar{z}/\sigma_Z$, is limited. Some indications, albeit approximate, may be derived from the following considerations. If the limit state function

$$z = g(x_1, x_2, \ldots, x_n)$$

is approximately linear, then the central limit theorem of probability calculation suggests that the distribution of X should tend toward the normal distribution. In that case, with reference to (2.29), an estimate of p_f is given by:

$$p_f = \Phi(-\beta) .$$

Moreover, in more general terms, some of the standardised cumulative distributions that are more commonly used in the field of structural safety exhibit tails which can be approximated by an exponential function of the type:

$$F_u(u) \cong A e^{4,6 u} \quad \text{(u being negative)} \tag{3.27}$$

Hence, setting:

$$u = (z - \bar{z})/\sigma_Z ,$$

$$p_f = P\left\{Z < 0\right\} = F_z(z=0) = F_u(u = -\bar{z}/\sigma_Z) = F_u(u = -\beta)$$

then, on the basis of (3.27):

$$p_f = A e^{-4,6 \beta} . \tag{3.28}$$

Increasing β by $1/2$, the value p_f derived from (3.28) is multiplied by the factor:

$$e^{-2,3} = 1/10 .$$

These considerations may serve as reference for the calibration of the values of β in a code procedure of the "second moment" type. For instance, Rosenblueth and Esteva [9] suggested that the functional relationship between β and p_f might be expressed in the form:

$$p_f = 460 e^{-4,3 \beta} , \tag{3.29}$$

which is an expression analogous to (3.28).

3.2.1.3 Invariance considerations

Owing to its operational simplicity, the MFOSM method undoubtedly has advantages in practice. The main criticism against it, which was probably the reason why other forms of the second moment theory have been advanced, is the following.

If we denote the two output variables by R and S, the failure criterion $R < S$ may be written in several forms, such as:

$$z = r - s < 0 \qquad \text{(a)}$$
$$z = (r - s)^3 < 0 \quad \text{(b)}$$
$$z = (r/s) - 1 < 0 \quad \text{(c)}$$
$$z = \ln (r/s) < 0 \qquad \text{(d)}$$

and various others.

Among these forms, (a) and (b) are equivalent, insofar as they define the same safety domain, in the plane r, s, for either positive or negative values of the variables. All four criteria (a), (b), (c) and (d) are equivalent provided that R and S are positive variables (Fig. 3.1). Now, since in any case the safety index is expressed in the form:

$$\beta = \bar{z}/\sigma_Z \quad ,$$

calculating the mean value and standard deviation of Z according to MFOSM operational schemes, the values of β obtained from the various criteria are not necessarily the same.

The same situation is encountered in the presence of unexceptionable mathematical transformations related to the definition of the resistances and of the load effects in the context of a predetermined failure criterion [32]. Hence, the method lacks invariance for transformations of the failure criterion which are consistent with the laws of algebra and mechanics (in addition to dimensional invariance, see Section 3.3).

Similarly, the problem may be shown to exist when operating in the space of the basic variables. Linearisation (3.17) corresponding to the mean values of the basic variables may yield different z' functions (3.18) for different, although equivalent, failure criteria, and therefore such as do not give the same value of $\beta = \bar{z}'/\sigma_Z'$, the first two moments of the variables X_i being equal. It is obvious that in these cases even the boundaries $z' = 0$, corresponding to limit state functions which are equivalent and linearised in the neighborhood of the mean values of the variables, will not be coincident.

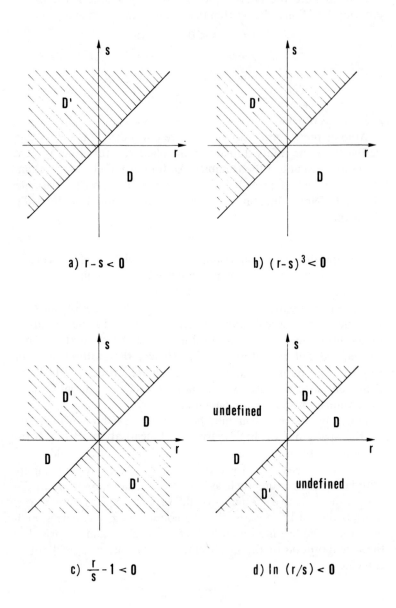

a) $r - s < 0$

b) $(r - s)^3 < 0$

c) $\dfrac{r}{s} - 1 < 0$

d) $\ln (r/s) < 0$

Fig. 3.1   From [35]

For example, Fig. 3.2 shows the limit state function:

$$z = g(r, s) = r - s \qquad \text{(a)}$$

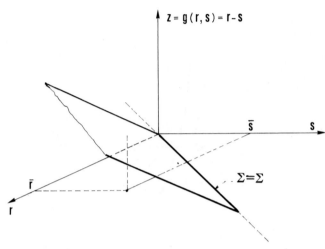

Fig. 3.2

and Fig.3.3 the equivalent function:

$$z = g(r, s) = (r - s)^3 \quad . \qquad \text{(b)}$$

The linearisation of (b) at the point having coordinates $\bar{r}$, $\bar{s}$, defines a boundary Σ', corresponding to:

$$z' = g'(r, s) = 0$$

and different from:

$$r - s = 0$$

which characterises (a).

As is shown below (Section 3.2.2.1.1), the distance of (3.18) projected into the space of the standardised variables from the origin is the MFOSM reliability index. Hence, it is understandable that in the evolution of the second moment theory the following strategy was adopted: linearisation, if necessary, of the limit state function $z = g(x_1, x_2, \ldots, x_n)$ not in correspondence to the mean values $\bar{x}_1, \bar{x}_2, \ldots, \bar{x}_n$, but at a point of the boundary $g(x_1, x_2, \ldots, x_n) = 0$, identical for all equivalent failure criteria.

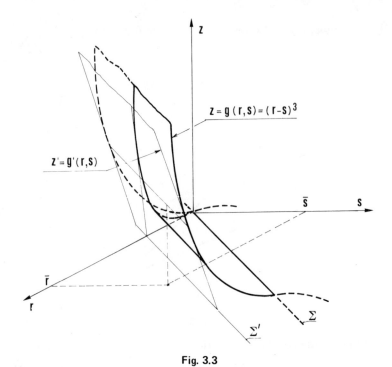

Fig. 3.3

3.2.2 The Lind-Hasofer "Minimum Distance" Method

Let $\underline{X}[X_1, X_2, \ldots, X_n]$ be the vector of the basic random variables, assumed to be uncorrelated, involved in a given structural problem. Their statistical character will be described by their mean values and variances. If there were correlation between the X_i expressed in terms of a known covariance matrix, it would always be possible to replace these variables by uncorrelated ones by means of the orthogonal transformation (2.14).

This operation does indeed make the problem more complicated from the analytical viewpoint; however, within approximations that are compatible with those of level 2 methods, if the partial correlation between two given variables X_i and X_j is weak (for instance, $\rho_{X_i X_j} < 0.2$), then it can be neglected; on the other hand, if it is high (for instance $\rho_{X_i X_j} > 0.8$), then these variables can be assumed to be exactly dependent; this last operation reduces the dimensions of the problem by unity.

Let:

$$z = g(x_1, x_2, \ldots, x_n) = 0 \qquad (3.30)$$

be the boundary of the safety domain. The values of $\underline{X}$ belonging to the failure domain will satisfy the inequality:

$$z = g(\underline{x}) < 0 \ .$$

The Lind-Hasofer method [33] consists of projecting (3.30) in the space of the standardised variables:

$$u_i = \frac{x_i - \bar{x}_i}{\sigma_{X_i}} \tag{3.31}$$

and of measuring, in this space, the minimum distance β of the transformed surface:

$$g(u_1, u_2, \ldots, u_n) = 0 \tag{3.32}$$

from the origin of the axes. A design is regarded as reliable, at the β^* level, prescribed by an appropriate code provision, if $\beta \geqslant \beta^*$. That is, the hypersphere having radius β^*, and with its centre at the origin of the axes u_i (corresponding to the mean values of the variables X_i), is required to lie entirely in the transformed safety domain.

The method is justified as follows: it is assumed that most, or at least an appropriate fraction, of the joint probability density of the variables involved will be concentrated in the hypersphere having radius β^* and that consequently it will be associated with values of vector $\underline{X}$ belonging to the safety domain. Moreover, the use of standardised variables ensures that the measurements along the different directions in the space under consideration will be nondimensional and comparable.

From the operational viewpoint the following minimisation problem must be resolved:

$$\left\{ \begin{array}{l} \beta = \min \sqrt{\sum_{i=1}^{n} u_i^2} \\[2em] g(u_1, u_2, \ldots, u_n) = 0 \end{array} \right. \tag{3.33}$$

3.2.2.1 Linear boundaries of the safety domain

Where the safety domain boundary is linear, for instance, if it corresponds to the equation:

$$z = g(x_1, x_2, \ldots, x_n) = b + \sum_{i=1}^{n} a_i x_i = 0 \ , \qquad (3.34)$$

β can be immediately determined. In fact, taking (3.31) into account, (3.34) becomes:

$$g(u_1, u_2, \ldots, u_n) = b + \sum_{i=1}^{n} a_i \bar{x}_i + \sum_{i=1}^{n} a_i \sigma_{x_i} u_i = 0$$

and the distance of this hyperplane from the origin of the axes is given by:

$$\beta = \frac{\displaystyle\sum_{i=1}^{n} a_i \bar{x}_i + b}{\sqrt{\displaystyle\sum_{i=1}^{n} a_i^2 \sigma_{x_i}^2}} \qquad (3.35)$$

3.2.2.1.1 Geometric interpretation of the MFOSM reliability index

Linearising (3.17) in accordance with the mean values of the basic variables, that is replacing (3.17) by (3.18), the Lind-Hasofer method would give the same results as those obtained by the MFOSM. The safety domain boundary Σ corresponding to (3.18) would in fact be expressed, in the space of the standardised variables, by the equation:

$$g(\bar{x}_1, \bar{x}_2, \ldots, \bar{x}_n) + \sum_{i=1}^{n} \left(\frac{\partial g}{\partial x_i}\right)_{\bar{x}} \sigma_{x_i} u_i = 0 \ ,$$

and the distance of the hyperplane from the origin of the axes would be:

$$\beta = \frac{g(\bar{x}_1, \bar{x}_2, \ldots, \bar{x}_n)}{\sqrt{\displaystyle\sum_{i=1}^{n} \left[\left(\frac{\partial g}{\partial x_i}\right)_{\bar{x}} \sigma_{x_i}\right]^2}} \ ,$$

This is coincident with (3.19) and is therefore characterised by the anomalies pointed out in 3.2.1.3. Hence, for limit state surfaces of any form, the problem (3.33) must be solved and this may involve considerable analytical difficulties.

3.2.2.2 Nonlinear boundaries of the safety domain

The assumption of the minimum distance β, obtained from a solution of problem (3.33), as a measure of reliability, is equivalent to the discretisation at one point only of the safety domain boundary, expressed in the space of the standardised variables. This corresponds to the substitution of the hypersurface by the hyperplane passing through P* and normal to OP*, O being the origin of the axes u_i and P* the point of minimum distance, referred to in technical literature as the "design point" (Fig. 3.4 illustrates the two-dimensional case).

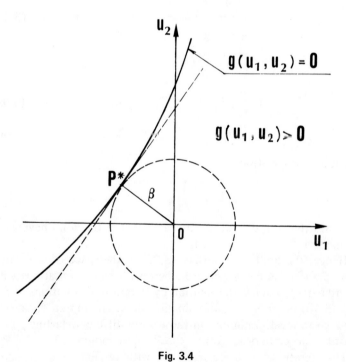

Fig. 3.4

Bearing in mind that, for a reasonable design, the point having coordinates $u_1=0$, $u_2=0$, . . . , $u_n=0$ belongs to the safety domain; that is:

$$g(0, 0, \ldots, 0) > 0 \tag{3.36}$$

let u_1^*, u_2^*, . . . , u_n^* be the coordinates of P* and β be the distance OP*. If (3.32) is capable of differentiation in the neighborhood of P*, the following considerations apply. Denoting by a_i the direction cosines of the straight line OP* oriented toward the failure region and normal to the limit state surface, considering that they are proportional to the partial derivatives $(\partial g/\partial u_i)$ calculated at P*, and that the condition:

$$\sum_{i=1}^{n} a_i^2 = 1 \qquad (3.37)$$

is satisfied, then:

$$a_i = -\frac{\left(\dfrac{\partial g}{\partial u_i}\right)_{\underline{u}^*}}{\sqrt{\displaystyle\sum_{j=1}^{n}\left(\dfrac{\partial g}{\partial u_j}\right)_{\underline{u}^*}^{2}}} \qquad (i = 1, 2, \ldots, n) \qquad (3.38)$$

$$u_i^* = \beta a_i \quad (i = 1, 2, \ldots, n) \qquad (3.39)$$

$$g(a_1\beta, a_2\beta, \ldots, a_n\beta) = 0 \qquad (3.40)$$

In (3.38) the notation

$$\left(\frac{\partial g}{\partial u_i}\right)_{\underline{u}^*}$$

indicates that the derivatives are calculated at the point having coordinates u_i^* (i = 1, 2, . . . , n).

Hence, the limit state surface (3.32) in the space of the standardised variables being known, albeit with some limitations with regard to its geometry, the solution by iteration of equations (3.38) and (3.40) permits the determination of the minimum distance β (or a given local minimum; in those cases different starting points must be chosen in order to find the absolute minimum).

The iteration can be carried out with respect to the original variable X_i instead of the standardised variables U_i. On the basis of (3.31) and (3.39), the coordinates of the design point in the space of the variables X_i are:

$$x_i^* = u_i^* \sigma_{X_i} + \bar{x}_i = \bar{x}_i + a_i \beta \sigma_{X_i} . \qquad (3.41)$$

Since:

$$\left(\frac{\partial g}{\partial u_i}\right)_{\underline{u}*} = \left(\frac{\partial g}{\partial x_i}\right)_{\underline{x}*} \sigma_{X_i} \qquad (3.42)$$

and the expression of the direction cosines a_i in reference to x_1, $x_2, \ldots, x_n$ will be:

$$a_i = -\frac{\left(\frac{\partial g}{\partial x_i}\right)_{\underline{x}*} \sigma_{X_i}}{\sqrt{\displaystyle\sum_{j=1}^{n}\left[\left(\frac{\partial g}{\partial x_j}\right)_{\underline{x}*} \sigma_{X_j}\right]^2}} \qquad (3.43)$$

Consequently, in this case, system (3.43) must be solved together with the equation:

$$g(\bar{x}_1 + a_1\beta\sigma_{X_1}, \ldots, \bar{x}_n + a_n\beta\sigma_{X_n}) = 0 \ . \qquad (3.44)$$

From the operational viewpoint one may start out by setting $x_1{}^* = \bar{x}_1$, $x_2{}^* = \bar{x}_2, \ldots, x_n{}^* = \bar{x}_n$, followed by the evaluation of the a_i factors by means of (3.43), by the determination of the value of β which satsifies (3.44), and by the calculation of the components of $x_i{}^*$ on the basis of (3.41) successively. These operations must be repeated until, in two successive iterations, the differences between the values of β and of $x_i{}^*$ respectively are found to be lower than predetermined quantities, that is, until one arrives at values for which (3.41), (3.43), and (3.44) are all valid simultaneously. Generally, the parameter β cannot be made explicit in (3.44) and one must resort to appropriate procedures (e.g., dichotomic methods) to determine the roots of this equation.

If instead of a verification problem (to determine β knowing all the parameters that appear in (3.30)), the task is to solve a design problem (e.g. to determine an unknown deterministic ϑ parameter with a target safety level β), then system (3.43) must be solved together with the equation:

$$g(\vartheta, x_1{}^*, x_2{}^*, \ldots, x_n{}^*) = 0 \ , \qquad (3.45)$$

The procedure could consist of the following operations:

1. fix a value of ϑ,
2. estimate point $\underline{x}^*$(intially one can set: $x_i^* = \bar{x}_i$ $(i = 1, 2, .., n))$,
3. compute a_i on the basis of (3.43),
4. calculate $x_i^* = \bar{x}_i + a_i\beta\sigma_{x_i}$,
5. determine ϑ by solving (3.45),
6. repeat steps 3 to 5 until stable value of ϑ and of $\underline{x}^*$ are obtained.

One can operate in a similar manner if the unknown quantity of the design is a parameter of the distribution of a statistical variable (for instance, $\bar{x}_1$).

Should the safety domain boundary turn out to be particularly complex and/or not capable of differentiation, specific mathematical procedures must be used to solve the minimisation problem (3.33) (see e.g.: Powell,M.J.D. "An efficient method for finding the minimum of a function of several variables without calculating derivates", *Computer Journal*, 7, 1964), or alternatively the method described in Section 3.2.2 can be adopted.

3.2.3 Considerations On Nonparametric Second Moment Methods

The concepts upon which the Lind-Hasofer method is based made a substantial contribution to the evolution of the second moment theory and inspired, more or less directly, all the proposals that followed, even those in the parametric field. In particular, this method, as opposed to MFOSM, is invariant in respect of different equivalent formulations of the failure criterion, in spaces having the same dimension. Unfortunately, in spite of the clear definition, the minimum distance is by no means easy to find for safety domain boundaries of any given form, especially if they present peaks and local minima. Furthermore, the method does not take full account of the geometry of these surfaces, as it considers only their point of contact with the hypersphere centred at the origin of the axes of the standardised variables. Consequently, the reliability information contained in β is limited and, since no assumption is made as to the distributions of the variables, it can be deduced only within the inequalities of Tchebychef. This information also depends (see 3.3) on the dimensions of the space in which β is defined. Therefore, even the Lind-Hasofer method lacks dimensional invariance, insofar as the same β value is associated with different lower and upper reliability bounds depending upon the dimension n of the space of the standardised variables in which one operates.

Thus, unless one imposes severe restrictions or makes appropriate classifications of problems, the safety indices that can be deduced from nonparametric second moment methods have little significance in terms of reliability. This situation can be improved by abandoning the nonparametric formulation of the problem.

3.3 METHODS THAT MAKE USE OF ASSUMPTIONS ON DISTRIBUTIONS

As seen in 3.1 and 3.2, in nonparametric second moment methods the variable quantities are expressed only in terms of their first two moments, no assumption being made about the type of their distributions. Owing to such a partial representation of the random variables, the probability of attaining the limit state is left, in a broad sense, undetermined. It is true that, theoretically, upper and lower bounds within which reliability lies could be deduced as Tchebychef inequalities if the Lind-Hasofer β index and the dimension n of the operating space are known [34]. Unfortunately, however, in the multidimensional field these inequalities are known only for some particular shapes of the safety domain boundary. Furthermore, if n turns out to be high, the bounds within which the reliability is comprised will be very wide. Another difficulty is encountered in connection with inference problems [35]. On the other hand, some information concerning reliability can be more easily deduced if the distribution functions of the random variables are known. It is obvious that in this case, all other conditions being equal, the bounds on p_f will be narrower than in the nonparametric formulation. Furthermore, in the case of multinormal distribution of the random variables and for particularly simple safety domain boundaries, to given values of β and of n, there will be corresponding exact values of p_f.

These considerations, added to the fact that reasonable estimates of p_f are required in formulating optimisation problems, lead to the conclusion that it is of little use to disregard the available information, be it limited, on the distributions of the random variables involved.

This gave rise to the formulation of level 2 methods that make use of distributional assumptions, while still operating within the basic principles of second moment methods: taking into account only the means, variances, and covariances of the random variables, checking safety at one (or more than one) point of the failure domain boundary.

3.3.1 Multinormal Distribution of the Basic Variables

One of the references that is most frequently used and is of considerable interest from the operational viewpoint is that of taking into account the normal distributions of the random variables. In this connection, let:

$$z = g(u_1, u_2, \ldots, u_n) < 0 \qquad (3.46)$$

be the failure criterion, in the space of the normal, independent and standardised variables U_i. It is possible to distinguish two specific configurations of (3.46) for which the Lind-Hasofer index $\beta \cdot$ has a fully defined probabilistic interpretation.

1st case - The limit state surface is linear.

Equation (3.46) can be expressed in the form:

$$z = b - \sum_{i=1}^{n} a_i u_i < 0 \ , \ b > 0 \qquad (3.47)$$

If:

$$D = \sqrt{\sum_{i=1}^{n} a_i^2} \ ,$$

then:

$$\beta = b/D$$

and the safety domain boundary is:

$$\beta - \sum_{i=1}^{n} \frac{a_i u_i}{D} = 0 \qquad (3.48)$$

Denoting by W the gaussian variable defined by:

$$w = (a_1 u_1 + a_2 u_2 + \ldots + a_n u_n)/D , \qquad (3.49)$$

(3.48) becomes:

$$\beta - w = 0 .$$

Hence the probability of failure is given by:

$$p_{f_L} = P\left\{(\beta - w) < 0\right\} = P\left\{w \geqslant \beta\right\} = 1 - \Phi(\beta) . \qquad (3.50)$$

In this case p_{f_L} does not depend on n.

2nd case - The limit state surface is a hypersphere centered at the origin with radius β.

The equation of this surface, always in the space of the standardised variables, will then be:

$$z = g(u_1, u_2, \ldots, u_n) = \beta^2 - \sum_{i=1}^{n} u_i^2 = 0 \qquad (3.51)$$

Bearing in mind that the random variable V defined by :

$$v = \sum_{i=1}^{n} u_i^2 , \qquad (3.52)$$

being the sum of the squares of n normal, standardised, and independent variables, has Chi-square distribution with n degrees of freedom, the failure probability is given by:

$$p_{f_S} = P\left\{(\beta^2 - v) < 0\right\} = P\left\{v \geqslant \beta^2\right\} = 1 - \chi_n^2(\beta^2) , \qquad (3.53)$$

where $\chi_n^2(\cdot)$ represents the cumulative distribution function Chi-square with n degrees of freedom.

Fig. 3.5 shows the functional relation (3.53) for n = 1, 2, 3, 4, 5. It is seen that, β being equal, p_{f_S} increases noticeably as the number n of the random variables increases. In the same diagram, the curve drawn with dashes represents the relation (3.50) which is independent of n.

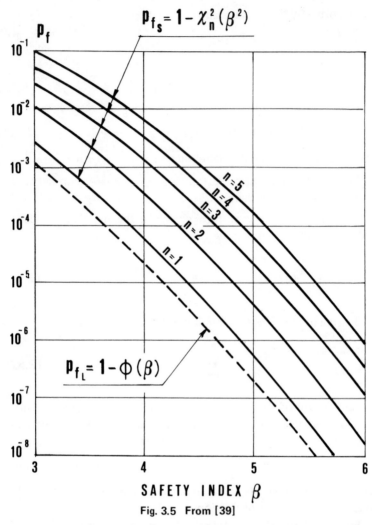

$$p_{f_s} = 1 - \chi_n^2(\beta^2)$$

$$p_{f_L} = 1 - \Phi(\beta)$$

SAFETY INDEX β

Fig. 3.5 From [39]

From the analysis of the two cases examined above, it follows that the minimum distance β can be interpreted probabilistically only by considering the geometry of the limit state surface and the dimensions of the operating space. Keeping in mind the concept of safety domain and (2.4), always referring to normal standardised variables, the following conclusions can be drawn:

a) if the limit state surface is linear and has distance β from the origin (Fig. 3.6), then, on the basis of (3.50):

$$p_f = p_{f_L} = 1 - \Phi(\beta)$$

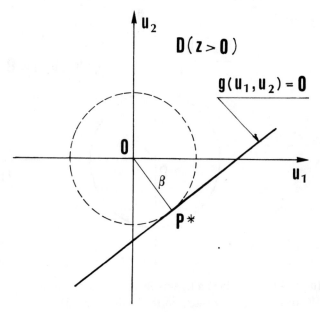

Fig. 3.6

b) if the safety domain is concave (Fig. 3.7), then:

$$0 \leqslant p_f \leqslant p_{f_L} \qquad (3.54)$$

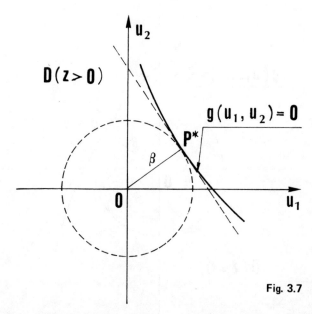

Fig. 3.7

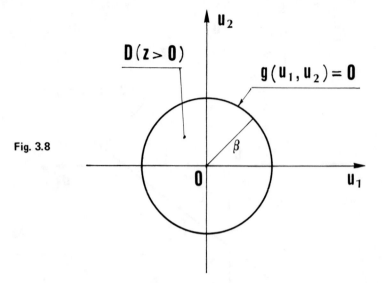

Fig. 3.8

c) if the safety domain is a hypersphere having radius β and center at the origin (Fig. 3.8) then, on the basis of (3.53):

$$p_f = p_{f_S} = 1 - \chi_n^2 \, (\beta^2)$$

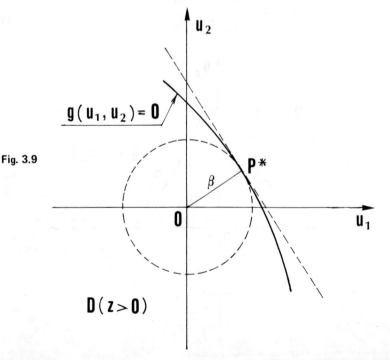

Fig. 3.9

d) if the safety domain is convex (Fig. 3.9), then:

$$p_{f_L} \leqslant p_f \leqslant p_{f_S} \qquad (3.55)$$

If attention is limited to special classes of safety domains, narrower upper and lower reliability bounds can be deduced [36].

In some structural problems the limit state surface does not deviate much from linear behaviour (and in that case (3.50) gives an acceptable estimate of failure probability). However, the generalisation of this position is not quite valid, since often one has to deal with markedly nonlinear limit state surfaces, or with limit state surfaces having more than one local minimum with respect to the distance function, or else that are not capable of differentiation in the neighbourhood of the point of minimum distance. In these cases, and particularly when the safety domain is defined in implicit form, the method consisting of the discretisation at several points of the safety domain boundary can be very useful (3.3.2).

3.3.2 Veneziano's Multiple Checks Method

With Veneziano's multiple checks method, the intersections of the safety domain boundary expressed in the space of uncorrelated and standardised variables, with a number of straight lines starting from the origin, is determined. A possible choice can be, for instance, that of the 2n directions of the axes and the 2^n directions of the bisecting lines (having director cosines:

$$a_i = \pm\, n^{-1/2} \qquad (i = 1, 2, \ldots, n)\ ),$$

which globally define $2n + 2^n$ intersection points and the same number of β_i distances of these points from the origin. The safety domain boundaries can be approximately replaced by spherical sectors (circular sectors, in the simple case shown in Fig. 3.10).

If the variables U_i have normal distribution, with reference to (3.53) and giving equal weight to all the distances β_i, an estimate of the failure probability is given by the relation:

$$p_{f_V} = \frac{1}{2n + 2^n} \sum_{i=1}^{2n+2^n} [1 - \chi_n^2\,(\beta_i^2)] \qquad (3.56)$$

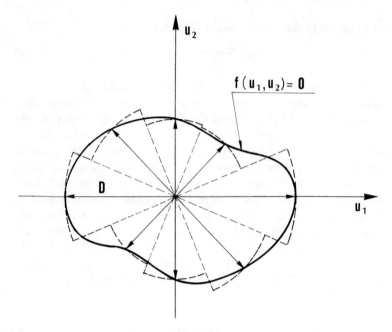

Fig. 3.10

If β_i = const. = β, then:

$$p_{f_V} = p_{f_S} = p_f \ .$$

Approximate values of the Lind-Hasofer index β_{min} and of the relative failure probability p_{f_L} can be deduced, without having to solve any minimisation problem, from the relations:

$$\beta_{min} \cong \min[\beta_i] \qquad (3.57)$$

$$p_{f_L} \cong 1 - \Phi(\beta_{min}) \qquad (3.58)$$

Moreover, from a number of examples reported by Veneziano [37], it seems that a better estimate of p_f than that given by (3.56) can be obtained by means of the relationship:

$$p_f = \max [p_{f_V}, p_{f_L}] \qquad (3.59)$$

in which, if the Lind-Hasofer index is unknown, the value obtained on the basis of (3.57) and (3.58) can be taken for p_{f_L} .

In conclusion, one can say that the method examined above can be easily applied operationally as it does not require the solution of minimisation problems, is less sensitive to dimensional invariance, compared with other methods, and is shown to be particularly useful when the safety domain is markedly convex and/or in the presence of discontinuities of the first derivatives of the limit state equation.

3.4 THE TRANSFORMATION OF ANY GIVEN DISTRIBUTION INTO NORMAL DISTRIBUTION

Nearly all the attempts made so far to generalise the level 2 methods discussed above to any distribution of the random variables essentially amount to the transformation, either rigorous or approximate, of the variables that have non-normal distributions into normal standardised variables. Such operational procedures can be understood in the light of the remarks made in Section 3.3 on the possibility of the probabilisitic interpretation of the minimum distance β (or of an appropriate set of distances) of the safety domain boundary from the origin in the space of the standardised gaussian variables. For the case of safety checking at one point only, the relation (3.50):

$$p_f = p_{f_L} = \Phi(-\beta) \ ,$$

which holds for linear safety domain bounaries, is generalised, in current practice, to boundaries of any type within standardised operational procedures. It is obvious that with the value of p_f determined in this manner one cannot associate a probabilisitic meaning (in the sense of real frequencies of failure) but merely an "operational" one.

From a thoeretical standpoint, it is possible to replace a random variable X_i, having a given cumulative distribution function $F(x_i)$, by a normal standardised variable U_i. The two variables are found to be related by the equality of probabilities:

$$F(x_i) = \Phi(u_i) \ , \tag{3.60}$$

i.e.,

$$u_i = \Phi^{-1}[F(x_i)] \ , \tag{3.61}$$

$$x_i = F^{-1}[\Phi(u_i)] \ . \tag{3.62}$$

For instance, if X_i is a type I extreme variable, (3.60) becomes

$$e^{-e^{-1,2825(x_i-\bar{x}_i)/\sigma_{X_i}-0,5772}} = \frac{1}{\sqrt{2\pi}} \int_{-\infty}^{u_i} e^{-t^2/2} \, dt \ ,$$

and (3.62) becomes:

$$x_i = \left[\frac{-\ln[-\ln \Phi(u_i)] - 0,5772}{1,2825} \right] \sigma_{X_i} + \bar{x}_i$$

Transformation (3.62) is generally nonlinear and nonelementary. On the other hand, if X_i is log-normally distributed (2.8.2) it is easy to determine a simple approximate functional relationship. In this case, where:

$$y_i = \ln x_i \ ,$$

the random variable Y_i is normal, with a mean value:

$$\bar{y}_i = \ln \bar{x}_i - \frac{1}{2} \ln (c_{X_i}^2 + 1) \cong \ln \bar{x}_i \qquad (3.63)$$

and with variance:

$$\sigma_{Y_i}^2 = \ln (c_{X_i}^2 + 1) \cong c_{X_i}^2 \qquad (3.64)$$

After standardising the variable Y_i, it follows that:

$$u_i = \frac{y_i - \bar{y}_i}{\sigma_{Y_i}} \cong \frac{\ln x_i - \ln \bar{x}_i}{c_{X_i}} = \frac{1}{c_{X_i}} \ln \frac{x_i}{\bar{x}_i} \qquad (3.65)$$

and therefore:

$$x_i \cong \bar{x}_i \, e^{u_i c_{X_i}} \qquad (3.66)$$

Actually (3.63) and (3.64) are valid only for small values of c_{X_i}, and furthermore (3.66) should be used only for resisting variables in order to avoid numerical inconsistencies [38].

Once the standardised gaussian variables U_i are known, it is possible to transform the safety domain in this space and determine $\beta_{\min}$, or, if one so wishes, the $2n + 2^n$ distances β_j. A shortcoming of the method lies in the fact that the transformations (3.61) usually render nonlinear even a safety domain boundary which is linear in the space of the basic variables X_i. Consequently, in any case, it is only possible to deduce the lower and upper bounds within which the reliability is contained. This may justify the adoption of standardised operational models that make use, in any case, of direct assumptions of normality for the variables involved (or log-normality for the resisting variables which can be rendered normal by means of (3.65)) or that transform the given variables into gaussian ones by means of approximate transformations.

To this area belong the following procedures, which have been used in practice or were proposed for draft codes:

—the method adopted by the "Nordic Committee for Building Regulations" [27]. Although this method was proposed within nonparametric second moment procedures [38], the resulting formulae can be justified on the basis of the following positions: assumption of log-normal distributions for the variables unfavourable to safety when they take on small values, assumption of gaussian distribution for the variables unfavourable to safety when they take on large values, determination of the minimum distance β of the limit state surface boundary from the origin of the standardized variables on the basis of the iterative procedure illustrated in Section 3.2.2.2, calculation of the "operational" probabilities of failure by means of the relation:

$$p_f = \Phi(-\beta) \ .$$

—replacement of the variables X_i having a given cumulative distribution function $F(x_i)$ by normal variables having the same mean $\bar{x}_i$ and the same fractile of order p_f (if X_i is unfavourable to safety when it takes on small values) and of order $1 - p_f$ (if X_i is unfavourable to safety when it takes on large values). This method was proposed by Paloheimo-Hannus (see Section 3.4.1), and it was one of the first formulations of level 2 design to make use of distributional assumptions. In this case (3.60) is satisfied only in correspondence to the fractiles mentioned above.

—transformation of the basic variables X_i into gaussian variables by a Taylor expansion of (3.61), neglecting nonlinear terms, at the unknown design point. Determination of this point by means of iterative procedures analogous to those mentioned in 3.2.2.2 and substitution of the safety domain boundary by the tangent hyperplane. This method, to be described in 3.4.2, was suggested by Lind [39], formulated in terms of applications by Fiessler-Rackwitz [40], and proposed as a possible level 2 operational procedure in CEB bulletin n° 116 E [10].

3.4.1 The Paloheimo-Hannus Method

Let $X_1, X_2, \ldots, X_n$ be the basic random variables involved in a design problem; these are assumed to be stochastically independent and to have known cumulative distribution functions $F_1(x_1)$, $F_2(x_2), \ldots, F_n(x_n)$.

Let the failure criterion be written in the form:

$$z = g(\vartheta, x_1, x_2, \ldots, x_n) < 0 , \qquad (3.67)$$

where ϑ denotes an unknown parameter which must be so determined that the reliability of the structure is approximately equal to the target value $1 - p_f$.

This parameter can be determined by solving by means of iterative procedures the following equations with respect to ϑ:

$$g(\vartheta, x_1^*, x_2^*, \ldots, x_n^*) = 0 \qquad (3.68)$$

$$x_i^* = \begin{cases} \bar{x}_i + \beta_i^- a_i \sigma_i, \text{ with } F_i(\bar{x}_i - \beta_i^- \sigma_i) = p_f \text{ if } \left(\dfrac{\partial g}{\partial x_i}\right)_{\underline{x}^*} > 0 \\[3ex] \bar{x}_i + \beta_i^+ a_i \sigma_i, \text{ with } F_i(\bar{x}_i + \beta_i^+ \sigma_i) = 1 - p_f \text{ if } \left(\dfrac{\partial g}{\partial x_i}\right)_{\underline{x}^*} < 0 \end{cases} \qquad (3.69)$$

$$a_i = \frac{-\sigma_i \beta_i^{\pm} \left(\dfrac{\partial g}{\partial x_i}\right)_{\underline{x}^*}}{\sqrt{\displaystyle\sum_{j=1}^{n} \left[\left(\dfrac{\partial g}{\partial x_j}\right)_{\underline{x}^*} \beta_j^{\pm} \sigma_j\right]^2}} \qquad (3.70)$$

$$\sum_{i=1}^{n} a_i^2 = 1 \qquad (3.71)$$

In (3.69) and (3.70) the notation:

$$\left(\frac{\partial g}{\partial x_i}\right)_{\underline{x}^*}$$

indicates that the derivatives are calculated at the design point $\underline{x}^*$ of coordinates $x_1^*, x_2^*, \ldots, x_n^*$. The quantities β_i^- and β_i^+ represent the distances of the fractiles of order p_f and $1 - p_f$, respectively, from the means $\bar{x}_i$, expressed in terms of standard deviations σ_i. These quantities are positive and depend upon the distributions of the variables under consideration. As to the types of distribution which occur more frequently in practice, the following may be distinguished:

a) normal distribution:

$$\beta_i^- = \beta_i^+ = \beta = \Phi^{-1}(1 - p_f) \tag{3.72}$$

For $p_f = 10^{-5}$ one gets $\beta = 4.265$.

b) log-normal distribution:

$$\beta_i^- = \frac{1 - \exp[\Phi^{-1}(p_f)\sqrt{k} - k/2]}{c_{x_i}} \quad ,$$

$$\beta_i^+ = \frac{\exp[\Phi^{-1}(1 - p_f)\sqrt{k} - k/2] - 1}{c_{x_i}} \quad , \tag{3.73}$$

where:

$$k = \ln(c_{x_i}^2 + 1)$$

For $p_f = 10^{-5}$ and $c_{x_i} = 0.30$ it follows that:

$$\beta_i^- = 2.420 \quad ; \quad \beta_i^+ = 7.833 \quad .$$

c) type I extreme distribution:

$$\beta_i^- = \frac{\sqrt{6}}{\pi} \left\{ \ln[-\ln(p_f)] + 0.5772 \right\}$$

$$\tag{3.74}$$

$$\beta_i^+ = -\frac{\sqrt{6}}{\pi} \left\{ \ln[-\ln(1 - p_f)] + 0.5772 \right\}$$

For $p_f = 10^{-5}$ it follows that:

$$\beta_i^- = 2.355 \quad , \quad \beta_i^+ = 8.526 \quad .$$

With the aim of including this method within the system illustrated in Section 3.4, the equation (3.68) to (3.71) are justified
below by different considerations from those undertaken by
Paloheimo-Hannus [41,42].

Let X_i be the given basic random variable having cumulative distribution function $F_i(x_i)$, mean $\bar{x}_i$ and standard deviation σ_i. Let
Y_i be a gaussian variable having the same mean $\bar{y}_i = \bar{x}_i$ of X_i and
the same fractile of order p_f, or $(1 - p_f)$, depending on whether
$(\partial g/\partial x_i)_{\underline{x}*}$ is greater or smaller than 0.

Denoting the standard deviation of Y_i by σ_i', the equality of the
above fractiles entails the validity of the relations:

$$
\begin{cases}
\bar{x}_i - \beta_i^- \sigma_i = \bar{x}_i - \beta \sigma_i' & \text{for} \left(\dfrac{\partial g}{\partial x_i}\right)_{\underline{x}*} > 0 \ , \\[4mm]
\bar{x}_i + \beta_i^+ \sigma_i = \bar{x}_i + \beta \sigma_i' & \text{for} \left(\dfrac{\partial g}{\partial x_i}\right)_{\underline{x}*} < 0 \ .
\end{cases}
\tag{3.75}
$$

As a result, the standard deviation of Y_i is given by:

$$
\sigma_i' = \frac{\beta_i^{\pm} \sigma_i}{\beta}
\tag{3.76}
$$

Standardising Y_i it follows that:

$$
u_i = \frac{y_i - \bar{y}_i}{\sigma_i'} = \frac{y_i - \bar{x}_i}{\sigma_i'} = \frac{(y_i - \bar{x}_i)\beta}{\sigma_i \beta_i^{\pm}}
\tag{3.77}
$$

Clearly, with the approximation introduced, the transformation
(3.60) is valid only for fractiles of order p_f (or order $1 - p_f$). Hence,
in the hypothesis of replacing each variable $X_i[\bar{x}_i, \sigma_i]$ by the
gaussian variable $Y_i[\bar{x}_i, \sigma_i']$ within the approximations given above,
it is possible to work with the standardised variables U_i (3.77)
having multinormal distribution. In this space, the safety domain
boundary becomes:

$$
z = g(\vartheta, u_1, u_2, \ldots, u_n) = 0
\tag{3.78}
$$

Referring to the method illustrated in 3.2.2.2, let P* be the
point at minimum distance of (3.78) from the origin of the axes 0
for a ϑ value such that:

$$
0P* = \beta = -\Phi^{-1}(p_f) \ .
$$

Denoting the direction cosines of the straight line OP* oriented toward the failure domain by a_i, if (3.78) can be differentiated at P*, one gets:

$$u_i{}^* = \beta\, a_i$$

$$a_i = -\dfrac{\left(\dfrac{\partial g}{\partial u_i}\right)_{\underline{u}^*}}{\sqrt{\displaystyle\sum_{j=1}^{n}\left(\dfrac{\partial g}{\partial u_j}\right)_{\underline{u}^*}^{2}}}$$

$$\sum_{i=1}^{n} a_i{}^2 = 1$$

$$g(\vartheta, u_1{}^*, u_2{}^*, \ldots, u_n{}^*) = g(\vartheta, \beta\, a_1, \beta\, a_2, \ldots, \beta\, a_n) = 0$$

Thus, operating by means of iterative procedures with respect to ϑ it is possible to solve the given design problem. In the space of variables

$$y_i = \bar{x}_i + \sigma_i{}' u_i\ , \tag{3.79}$$

considering that.

$$\left(\dfrac{\partial g}{\partial u_i}\right)_{\underline{u}^*} = \left(\dfrac{\partial g}{\partial y_i}\right)_{\underline{y}^*} \sigma_i{}' \tag{3.80}$$

the design equation becomes:

$$g(\vartheta, y_1{}^*, y_2{}^*, \ldots, y_n{}^*) = 0\ , \tag{3.81}$$

with:

$$y_i{}^* = \bar{x}_i + \dfrac{\beta\, a_i \beta_i^{\pm}\sigma_i}{\beta} = \bar{x}_i + \beta_i^{\pm} a_i \sigma_i \tag{3.82}$$

where:

$$a_i = -\dfrac{\left(\dfrac{\partial g}{\partial y_i}\right)_{\underline{y}^*}\beta_i^{\pm}\sigma_i}{\sqrt{\displaystyle\sum_{j=1}^{n}\left[\left(\dfrac{\partial g}{\partial y_j}\right)_{\underline{y}^*}\sigma_j\beta_j^{\pm}\right]^{2}}} \tag{3.83}$$

$$\sum_{i=1}^{n} a_i^2 = 1 \qquad (3.84)$$

relations which are analogous with (3.68) to (3.71).

Obviously, where the random variables involved have multinormal joint distribution and the safety domain boundary is linear, the procedure under consideration gives exact results. In the other cases, approximate results are obtained.

In order to ascertain the accuracy of these approximations, in the context of a simple problem not requiring the use of iterative procedures and which has already been treated with other methods, let:

$$z = g\,(r, s) = r - s < 0$$

be the failure criterion, and let:

$$F_R\,(r)\,,\bar{r}\,,c_R \quad \text{and} \quad F_S\,(s)\,,\bar{s},c_S$$

be the cumulative distribution function, the means, and the coefficients of the variation of the two indpendent variables R and S. One wants to determine, for instance, $\bar{r}$, the other parameters being known. On the basis of the method under consideration, the design equation is:

$$g(r^*, s^*) = r^* - s^* = \bar{r} + a_R \beta_R^- \sigma_R - (\bar{s} + a_S \beta_S^+ \sigma_S) = 0 \qquad (3.85)$$

where:

$$a_R = - \frac{\sigma_R \beta_R^-}{\sqrt{\sigma_R^2 (\beta_R^-)^2 + \sigma_S^2 (\beta_S^+)^2}} \quad ,$$

$$a_S = \frac{\sigma_S \beta_S^+}{\sqrt{\sigma_R^2 (\beta_R^-)^2 + \sigma_S^2 (\beta_S^+)^2}} \quad ,$$

(3.85) becomes:

$$\bar{r} - \bar{s} = \sqrt{\sigma_R^2 (\beta_R^-)^2 + \sigma_S^2 (\beta_S^+)^2}$$

which makes it possible to solve the problem.

Introducing the central safety factor $\gamma_0 = \bar{r}/\bar{s}$, the equation becomes:

$$\gamma_0 - 1 = \sqrt{(\beta_R^-)^2 c_R^2 \gamma_0^2 + (\beta_S^+)^2 c_S^2}$$

and after squaring and rearranging:

$$\gamma_0 = \frac{1 + \sqrt{(\beta_R^-)^2 c_R^2 + (\beta_S^+)^2 c_S^2 - (\beta_R^-)^2 (\beta_S^+)^2 c_R^2 c_S^2}}{1 - (\beta_R^-)^2 c_R^2} \quad (3.86)$$

Expression (3.86) is then analogous with (2.37) if R and S are both normal. For distributions of R and S other than normal, one must operate numerically to be able to make comparisons. In Fig. 3.11 the results of this analysis are reported [40], considering various distribution functions of R and S and those values of the coefficients of variation c_R and c_S that are most frequently encountered in operational practice.

Fixing appropriate "operational" values p_f^* (in Fig. 3.11: $p_f^* = 10^{-3}, 10^{-5}, 10^{-7}$) and the distribution types of R and S (in Fig. 3.11: R:LN/S:EX I, R:LN/S:LN, R:N/S:LN, R:N/S:EX I), β_i^+ and β_i^- values can be deduced by means of (3.72) to (3.74).

On the basis of (3.86) one can calculate γ_0 as a function of c_R and c_S. Knowing these parameters, by solving the convolution integral (2.24) one obtains the actual failure probability p_f. These "operational" and "actual" failure probabilities are shown in Fig. 3.11. The various curves show an approximation which is acceptable for practical purposes. In the case of safety domain boundaries of any given form, the considerations in 3.3.1 are still qualitatively valid.

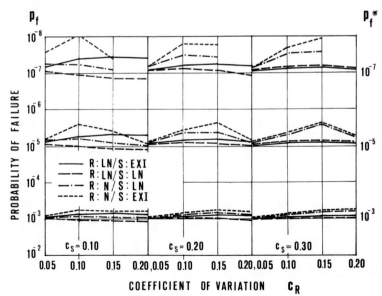

Fig. 3.11

3.4.2 The Method Proposed in the Appendix I to the CEB Information Bulletin n° 116–E [10]

Let $\underline{X}[X_1, X_2, \ldots, X_n]$ be the vector of the independent random variables involved in the structural problem, and let:

$$z = g(x_1, x_2, \ldots, x_n) = 0$$

be the safety domain boundary. The point having coordinates $x_1 = \bar{x}_1$, $x_2 = \bar{x}_2$, $\ldots$, $x_n = \bar{x}_n$ belongs to this domain, i.e., it follows that $g(\bar{x}_1, \bar{x}_2, \ldots, \bar{x}_n) > 0$. Let $F(x_i)$ be the cumulative distribution function of the variable X_i and

$$\underline{x}^*[x_1{}^*, x_2{}^*, \ldots, x_n{}^*]$$

the unknown design point. Furthermore, let Y_i be a gaussian variable and

$$u_i = (y_i - \bar{y}_i)/\sigma_{Y_i} \qquad (3.87)$$

the standardised variable concerned.

By a Taylor power series expansion of (3.61) in correspondence with $\underline{x}^*$ and considering linear terms only, one obtains:

$$u_i \cong \Phi^{-1}[F(x_i{}^*)] + \left\{ \frac{\partial \Phi^{-1}[F(x_i)]}{\partial x_i} \right\}_{\underline{x}^*} (x_i - x_i{}^*), \qquad (3.88)$$

where the notation:

$$\left\{ \frac{\partial \Phi^{-1}[\cdot]}{\partial x_i} \right\}_{\underline{x}^*}$$

indicates that the derivatives are calculated at the point having coordinates $x_1{}^*, x_2{}^*, \ldots, x_n{}^*$.

By the rules of derivation of the inverse function, (3.88) becomes:

$$u_i \cong \Phi^{-1}[F(x_i{}^*)] + \frac{f(x_i{}^*)(x_i - x_i{}^*)}{\varphi\left\{ \Phi^{-1}[F(x_i{}^*)] \right\}}, \qquad (3.89)$$

where $f(\cdot)$ and $\varphi(\cdot)$ denote the probability densities corresponding to the cumulative distributions $F(\cdot)$ and $\Phi(\cdot)$ respectively. Rearranging (3.89) one gets:

$$u_i \cong \frac{x_i - \left\{ x_i^* - \dfrac{\Phi^{-1}[F(x_i^*)] \, \varphi\left\{\Phi^{-1}[F(x_i^*)]\right\}}{f(x_i^*)} \right\}}{\dfrac{\varphi\left\{\Phi^{-1}[F(x_i^*)]\right\}}{f(x_i^*)}} \qquad (3.90)$$

(3.90) represents a linear transformation of the basic variable X_i into the standardised normal variable U_i, and it is equivalent, taking into account (3.87), to approximating the given distribution by a normal one having mean $\bar{y}_i$ and standard deviation σ_{Y_i} given by:

$$\bar{y}_i = x_i^* - \frac{\Phi^{-1}[F(x_i^*)] \, \varphi\left\{\Phi^{-1}[F(x_i^*)]\right\}}{f(x_i^*)} , \qquad (3.91)$$

$$\sigma_{Y_i} = \frac{\varphi\left\{\Phi^{-1}[F(x_i^*)]\right\}}{f(x_i^*)} . \qquad (3.92)$$

Hence the given variable $X_i[\bar{x}_i, \sigma_{X_i}]$ can be replaced, within the approximations effected, by the gaussian variable $Y_i[\bar{y}_i, \sigma_{Y_i}]$. Even if (3.61) is thus corrected only at the design point $\underline{x}^*$, the transformation effected gives good approximations in the proximity of this point, that is in the region of major interest for checking safety. This transformation is qualitatively represented in Fig. 3.12 on a normal probability paper, and it is equivalent to fitting any given distribution (solid line in Fig. 3.12) by a normal distribution (dashed line in Fig. 3.12), both having the same values of cumulative functions and also of probability densities at the design point [40].

Finally, if X_i is log-normal, it is more convenient to make use of (3.65) instead of (3.90). The design point having been initially fixed, once the standardised gaussian variables U_i have been obtained by means of (3.91), (3.92) and (3.87), the safety domain boundary can be expressed in the form:

$$g(u_1, u_2, \ldots, u_n) = 0 \qquad (3.93)$$

and the calculation of the minimum distance can be carried out, assuming that (3.93) is capable of differentiation, on the basis of the usual iterative procedures, that is by solving the equations:

$$g(u_1{}^*, u_2{}^*, \ldots, u_n{}^*) = g(\beta a_1, \beta a_2, \ldots, \beta a_n) = 0$$

$$a_i = -\left(\frac{\partial g}{\partial u_i}\right)_{\underline{u}^*} \Bigg/ \sqrt{\sum_{j=1}^{n}\left(\frac{\partial g}{\partial u_j}\right)^2_{\underline{u}^*}} \quad (i = 1, 2, \ldots, n)$$

The values $x_i{}^*$ and hence $\bar{y}_i$ and σ_{Y_i} must be recalculated in each iteration loop.

When a satisfactory value of the minimum distance β has been obtained, the "operational" failure probability p_f can be calculated on the basis of equation (3.50):

$$p_f = \Phi(-\beta) \ .$$

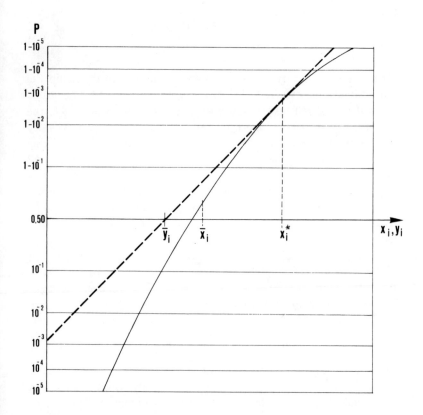

Fig. 3.12

This method is then characterised by first order approximations both in the normalisation of random variables (Taylor series expansion of (3.61) corresponding to the design point P* neglecting nonlinear terms) and in the calculation of failure probability (substitution at P* of the safety domain boundary by the tangent hyperplane, implicit in (3.50)).

3.4.3 Summarizing Considerations of the "Advanced" Level 2 Methods

From the above analysis of the "advanced" level 2 methods, with which safety checks are made at a finite number of points of the limit state surface, the following considerations arise:

a) It is useful to formulate the limit state equation in the space of the basic variables, in order to avoid approximations in determining the statistical parameters of the output variables.

b) As a measure of safety, it is useful to choose the minimum distance or an appropriate vector of distances of the safety domain boundary, expressed in the space of the standardised basic variables from the origin of the axes. These positions allow us to achieve measures of safety that are invariant at least with respect to different and equivalent formulations of the failure criterion in spaces having the same dimensions.

c) It is useful to choose methods that take into account the available information on the type of distributions of the random variables rather than nonparametric second moment procedures, while retaining the operational tools of the latter.

d) It is necessary to take into account the geometry of the safety domain boundary so as to be able to deduce information about the reliability of the design under consideration, and also, still within this interpretative framework, it is useful to transform any given type of distributions into gaussian ones, either by rigorous or approximate procedures. With these positions:

d_1) if the safety domain boundary in the space of the standardised variables is approximately linear, it is sufficient to check safety at the point of minimum distance from the origin. The relationship:

$$p_f = 1 - \Phi(\beta_{min})$$

gives the failure probability;

d_2) for safety regions that are concave and convex, (3.54) and (3.55) give generally quite ample bounds within which p_f lies;

d_3) especially in the case of convex safety domains, the multiple checking method 3.3.2 permits the deduction of values of p_f more significant than those that consider minimum distance only.

3.5 THE PROBLEM OF THE UNCERTAINTIES IN PROBABIL- ITIES AND MECHANICAL MODELS

In all safety analysis problems, assumptions and evaluations are made; these are characterised by uncertainties having different origins; as a first approximation, these can be grouped in two classes:

—uncertainties related to the type and parameters of the probabilistic model,

—uncertainties related to the mechanical model which forms the basis on which the limit state equation is formulated.

The first class comprises those unavoidable statistical uncertainties that originate from: incompleteness of the information, which can be deduced from the available data or from tests on samples of limited size in addition to the possible use of materials coming from different sources or manufactured on site under varying conditions.

As to the optimal treatment of uncertainty parameters, for instance in the first two moments, i.e., those which are taken into consideration for level 2 procedures, one may resort to Bayesian concepts and methodologies [18]. However, for small variabilities in the uncertainty parameters, it is possible to propose, for purposes of practical application, simpler and more approximate operational schemes. In this context, with every basic variable one can associate an uncertainty parameter X_{2i}, which is in turn regarded as an independent statistical variable, so that:

$$x_i = x_{1i} + x_{2i} \qquad (3.94)$$

or:

$$x_i = x_{1i} \cdot x_{2i} \qquad (3.95)$$

In the first case one has:

$$\bar{x}_i = \bar{x}_{1i} + \bar{x}_{2i}$$

$$\sigma^2_{X_i} = \sigma^2_{X_{1i}} + \sigma^2_{X_{2i}} \quad.$$

and in the second case:

$$\bar{x}_i = \bar{x}_{1i} \cdot \bar{x}_{2i}$$

$$c^2_{X_i} = c^2_{X_{1i}} + c^2_{X_{2i}}$$

Usually in the code proposals [10,27] the mean value $\bar{x}_{2i}$ is taken as equal to 0 in (3.94) and equal to 1 in (3.95), which simply gives rise to an increase in the scatter of the given basic variable.

For the purpose of taking account of the uncertainties, as well as the usual simplifications and approximations of the mechanical model (e.g., in the calculation of the action effects and resistances of the sections or, more generally, in the formulation of the limit state equation), it was proposed to introduce where appropriate one or more additional variables in the limit state equation [43,44]. Their distribution and relative parameters should be assessed by a comparison between laboratory tests and the results obtained from the chosen model.

3.6 MULTIPLE FAILURE MODES

If a structural problem is conditioned by the presence of k uncorrelated failure mechanisms and if each of these is characterised by its own limit state surface which is approximately linear in the space of the standardised normal variables, the determination of the overall failure probability becomes particularly easy. Taking into account (3.50) and the additive rule of the probability calculations in fact, one obtains:

$$p_f \cong \sum_{i=1}^{k} p_{f_i} \cong \sum_{i=1}^{k} [\Phi(-\beta_i)] \quad, \tag{3.96}$$

where β_i denotes the minimum distance from the origin of each of the k limit state surfaces. In any other case, the value obtained from (3.96) can be adopted as an "operational" value within the approximations given in 3.4. A valid alternative may at times consist of using (3.56) rather than (3.50), even in the problem under consideration.

3.7. THE COMBINATION OF LOADS VARYING WITH TIME

So far, all the random variables involved in a given structural problem have been assumed to be independent of time, that is, once they have taken on a given value, they maintain it unaltered over the entire expected lifetime of the structure. This may be true for the strengths of the materials and for permanent loads, but not for the remaining actions, since, depending on their nature, these will display a more or less marked variability with time. That means that these actions assume the character of stochastic processes, which are usually not very well known and are difficult to treat in analytical terms; a rigorous study of the topic belongs to the sphere of advanced reliability theory. In order to include them within the simpler and approximate methodologies pertaining to level 2, drastic simplifications must be effected. Among the various models that have been proposed for use in practical application, the one that is most commonly used is the "vertical classification" model developed by Ferry-Borges and Castanheta [22] within the framework of concepts proposed by Turkstra [45].

Operating in accordance with this scheme, for every time-varying action X_i the expected life span of the structure T is divided into r_i time intervals ΔT_i of equal length, referred to as elementary intervals. In each of these the action under study is regarded as constant and having a value equal to the maximum reached in it. Moreover it is assumed that ΔT_i is so chosen that the maximum values of the actions reached in successive intervals will be independent. Hence, denoting by $F(x_i)$ the cumulative distribution of the instantaneous X_i values, the cumulative distribution $F_{r_i}(x_i)$ of the maximum values of X_i in r_i elementary intervals, that is in the expected design life:

$$T = \Delta T_i \, r_i$$

fixed in advance for the structure, is:

$$F_{r_i}(x_i) = [F(x_i)]^{r_i} \ ,$$

and the probability density $f_{r_i}(x_i)$ is:

$$f_{r_i}(x_i) = r_i \, f(x_i) \, [F(x_i)]^{r_i - 1} \tag{3.97}$$

where $f(x_i)$ is the probability density of the instantaneous values of X_i.

For these assumptions, the joint probability density

$$f_{r_1, r_2, \ldots, r_k}(x_1, x_2, \ldots, x_k)$$

of k different actions characterised by: $r_1, r_2, \ldots, r_k$ independent repetitions (arranged in the increasing order of their respective values r_i (i=1,2,...,k)) is given by [22]:

$$f_{r_1, r_2, \ldots, r_k}(x_1, x_2, \ldots, x_k) dx_1 \, dx_2 \ldots dx_k =$$
$$1 - \left[1 - f(x_1)dx_1 \left\{ 1 - \left[1 - f(x_2)dx_2 \ldots \right. \right. \right.$$
$$\left. \left. \left. \ldots \left\{ 1 - [1 - f(x_k)dx_k]^{r_k/r_{k-1}} \right\} \ldots \right]^{r_2/r_1} \right\} \right]^{r_1} \quad (3.98)$$

and is then generally available only in numerical form. The introduction of approximations acceptable in conjunction with high fractiles of the distributions of the individual random variables involved, that is, for small values of $f(x_i)$ (which are, in any case, those of greatest interest for the actions), makes it possible to consider, instead of (3.98), k different joint distribution functions of particularly simple form, each of which gives the best fit to (3.98) in certain regions of the space of the variables X_i. In effect, a satisfactory approximation of the most critical joint action of the k time-varying loads can be ascertained by considering the combinations characterised by the following k probability densities:

$$f_{r_1, r_2, \ldots, r_k}(x_1, x_2, \ldots, x_k)$$

$$= \begin{cases} f_{r_1}(x_1) f_{r_2/r_1}(x_2) f_{r_3/r_2}(x_3) \ldots f_{r_k/r_{k-1}}(x_k) \\[6pt] f(x_1) f_{r_2}(x_2) f_{r_3/r_2}(x_3) \ldots f_{r_k/r_{k-1}}(x_k) \\[6pt] f(x_1) f(x_2) f(x_3) \ldots f_{r_k/r_{k-1}}(x_k) \qquad (3.99) \\[6pt] \cdots\cdots\cdots\cdots\cdots\cdots\cdots\cdots\cdots\cdots\cdots \\[6pt] f(x_1) f(x_2) f(x_3) \ldots f_{r_k}(x_k) \end{cases}$$

From an assessment of (3.99) and keeping in mind (2.13), it is seen that safety analysis in the presence of time-varying actions can be carried out by regarding these as ordinary independent random variables, provided that each of them is characterised by the cumulative distribution function $F_{r_i/r_j}(x_i)$ (and probability density $f_{r_i/r_j}(x_i)$, related to the former by (3.97)) which pertains to it in the given combination indicated in (3.99). The combination scheme suggested above therefore involves taking into account k different failure surfaces. The overall failure probability of the structure is then given by (3.94) within the validity of the positions stated in Section 3.6.

So far we have not considered explicitly the possibility that the action X_i may be intermittent. This problem has been dealt with, in different ways, by several research workers [46, 47, 10, . . .]. In this case the "as sampled" distribution $F(x_i)$ [21] of the instantaneous values X_i is characterised by a mass concentrated at the zero value expressing its probability of non-occurrence. If one wants to introduce the cumulative distribution function $F^*(x_i)$ of instantaneous values "given that the action X_i does occur", with the same idealisation of the actions considered so far and adding the following further conditions:

—the occurrence or nonoccurrence of the action X_i in each elementary time interval ΔT_i corresponds to repeated trials having probability p_i of success (occurrence),

—the occurrence and the intensity of the actions are independent,

then:

$$F(x_i) = \left\{1 - p_i \left[1 - F^*(x_i)\right]\right\} \qquad (3.100)$$

and equation (3.96) becomes:

$$F_{r_i}(x_i) = \left\{1 - p_i \left[1 - F^*(x_i)\right]\right\}^{r_i} \qquad (3.101)$$

Corresponding to the upper tails of the distributions of the actions, (3.101) can be approximated by the relationship:

$$F_{r_i}(x_i) \cong \left[F^*(x_i)\right]^{p_i r_i}$$

The cumulative distribution functions $F^*(x_i)$, their parameters, and the occurrence probabilities p_i (or the "as sampled" distributions $F(x_i)$) as well as the number of independent repetitions relative to a fixed reference period must be deduced on the basis of the observed statistical data for each individual action (see, for example, "Basic Notes on Actions" [21]).

Level 1 Methods

4.1 PREMISE

"Level 1" methods are defined as all those design and safety checking methods that take account of the statistical aspects of the problem by introducing in the limit state equations suitable nominal values of the basic variables, in association with partial safety factors. As far as possible, the latter are deduced from probabilistic considerations so as to ensure appropriate reliability levels. These nominal values are suitable parameters for the distributions of the basic variables, for instance, they can be the mean values, or more often the "characteristic values," these latter being conceived as fractiles of a predetermined order of the respective distribution functions, low for the mechnical material properties (e.g. 0.05) and high for the actions (e.g. 0.95).

Broadly, if the failure criterion in respect of a predetermined limit state is expressed in the form:

$$g(x_1, x_2, \ldots, x_h, x_{h+1}, x_{h+2}, \ldots, x_n) < 0 \ , \qquad (4.1)$$

safety analysis by means of level 1 methods is effected by checking that the following inequality is satisfied:

$$g(x_{1k}/\gamma_{m1}, x_{2k}/\gamma_{m2}, \ldots, x_{hk}/\gamma_{mh}, x_{(h+1)k}\gamma_{f(m+1)},$$
$$x_{(h+2)k}\gamma_{f(m+2)}, \ldots, x_{nk}\gamma_{fn}) \geqslant 0 \quad (4.2)$$

In (4.1) and (4.2):

X_r (r = 1, 2, . . . , h)   denote the resisting variables,

X_s (s = h+1, h+2, . . . , n) denote the actions,

$$x_{rk} = \bar{x}_r (1 - k_r c_{X_r}) \qquad (4.3)$$

are the characteristic values of the resisting variables,

$$x_{sk} = \bar{x}_s (1 + k_s c_{X_s}) \qquad (4.4)$$

the characteristic values of the actions, γ_{mr}, γ_{fs}, the partial safety factors.

Level 1 methods constitute the basis of present day codes that use probabilistic concepts. Adoption of the limit states methods and the use of partial safety factors were proposed on the international level by the Comité Euro-International du Béton (CEB) [48] and were later formally accepted in ISO/DIS 2394 [49]. Continuous revisions of the operational methods used [50, 10] are in progress and specific problems treated by the national codes of several countries are being reviewed. A number of alternative proposals for level 1 codes have been recently formulated by various research groups [27, 51, 52].

Proposals have also be made, in fact, for codes of practice based on the application of level 2 concepts as mentioned in Part 3. It appears that even the more advanced level 2 methods could be applied in codes of practice in the future, if agreement can be reached on the following issues:

−selection of basic random variables for each specific problem, their distribution types and relative statistical parameters,
−form of the various limit state equations and choice of models on the basis of which to include and deal with uncertainties,
−operational reliability levels to be adopted in different design situations.

Even if agreement could be reached among the various international organisations and in view of the indispensable requirement for knowledge and application of specific statistical methodologies and also because the solution of most problems requires calculation procedures involving the use of a computer, in principle the direct application of these methods is likely to be confined to special problems of particular economic and technical relevance, and to the evaluation of suitable sets of partial safety factors for level 1 methods.

Indeed, level 1 methods could be made identical to those of level of the parameter of the basic variables within a given limit state equation. To make it possible for level 1 methods to be applied to the solution of routine problems, however, it must be possible to treat a wide range of design situations by means of a given set of partial coefficients; hence the continuous functions mentioned above must be made into discrete elements. Such a discretisation can be carried out on the basis of various concepts and criteria. In theory, it is possible to choose different values for given partial safety factors, even trivial ones, for instance, unity. Moreover, practical and psychological reasons suggest a number of restrictions on the choice of these factors, viz.:

−the coefficients γ_m and γ_f in (4.1) should be greater than unity, at least for variables characterised by high sensitivity factors a_i (3.43),

−whenever possible the coefficients γ_f for the actions should be the same for structures built with different materials and of different types of construction,

−the design of structures which belong to different safety and control classes should be affected only by variation of the coefficients γ_m for the resisting variables.

Among the various methods that satisfy the requirements listed above, the advantages that distinguish level 1 schemes are their great operational simplicity owing to the use of fixed and constant partial safety factors for a given class of design situation; as distinct from levels 2 and 3, the probability of attaining a given limit state need not be directly calculated. On the other hand, the short-comings of the method derive from the fact that the reliability of individual designs will necessarily vary from case to case, that is, the objective of obtaining constant reliability cannot be reached for all the design situations covered in a given code.

One of the most important problems that must be faced in drafting level 1 codes is then the selection of partial safety factors for a given structural class in such a way that the efficiency of the method proposed is satisfactory, i.e., that the deviation of the reliability of a design made on the basis of the chosen coefficients, from the desired reliability level laid down in the code, should be acceptable. Principles and operational procedures for the practical evaluation of partial safety factors have been proposed in recent studies by Lind [53] and Baker [54] with the object of satisfying this condition.

4.2 THE SEMI−PROBABILISTIC CEP−FIP METHOD

The most widely known level 1 method which served as a basis for a number of national codes, especially in European countries, is that provided in the "International CEB-FIP Recommendations on the design and construction of concrete structures"[48]. With this method, safety analysis is carried out in the space of the output variables, comparing the effects of the actions with those of the material strengths. In reference to reinforced concrete structures and to the limit state equation:

$$g_R(f_c, f_s) - g_S(F_1, F_2, \ldots, F_n) = 0 \ , \qquad (4.5)$$

the safety checking relationship is written in the form:

$$g_R(f_{ck}/\gamma_{mc}, f_{sk}/\gamma_{ms}) - \gamma_{f3} g_S(\gamma_{f11} \gamma_{f21} F_{1k},$$

$$\gamma_{f12} \gamma_{f22} F_{2k}, \ldots, \gamma_{f1n} \gamma_{f2n} F_{nk}) \geqslant 0 \qquad (4.6)$$

In (4.5) and (4.6) the symbols used denote:

f_c concrete strength,

f_s steel strength,

F_i the given action applied to the structure,

$g_R(\cdot)$ the functional relation between the material strengths and their effects,

$g_S(\cdot)$ the functional relation between the actions and their effects,

$\left.\begin{array}{l} f_{ck}, \\ f_{sk}, \\ F_{ik} \end{array}\right\}$ the characteristic values of the material strengths and of the actions,

γ_m, γ_f the partial safety factors.

The characteristic values f_{ck} and f_{sk} are the fractiles of the order of 0.05 of the distributions of the material strengths. To determine these one can refer to normal distribution; if the estimates of the means and variances of the population were deduced from testing samples of small size, it is necessary to take due account of the uncertainties in these parameters [55]. The characteristic values F_{ik} are, generally, the fractiles of the order of 0.95 of the distributions of the actions. As to actions varying with time, the distributions used as reference are those of the maximum values in the expected design life of the structure.

With regard to the meaning of the partial safety factors, it can be stated that:

$-\gamma_{mc}, \gamma_{ms}$ reduce the characteristic values of the mechanical material properties and are intended to cover a number of uncertainties associated with these parameters (unfavourable deviations from the specified characteristic values, differences between the material strength in the structure and that resulting from specimen testing, possible local weaknesses in the structural material arising from or in the construction process, possible inaccurate assessment of the resistance of elements arising from the strength of the materials and manufacturing tolerances);

$-\gamma_{f1i}$ take account of possible unfavourable deviations of the actions from their characteristic values F_{ik},

$-\gamma_{f2i}$ are combination factors for the actions; they take account of the reduced probability of the simultaneous combined occurrence of the actions, all at their characteristic values;

$-\gamma_{f3}$ take account of the uncertainties in the assessment of the action effects and their significance for safety.

An additional coefficient γ_n may be introduced into (4.6) to take into account the type of structural failure involved (ductile or brittle), as well as the seriousness of the effect of attaining the limit state under consideration. This factor however should not be used explicitly, but should rather serve to modify γ_m or γ_f as the specific situation requires.

With regard to ultimate limit states and for the case of proportional relations between actions and load effects, the 1970 CEB-FIP Recommendations provide the following relationships for checking the safety of reinforced concrete structures:

$$g_R\left(\frac{f_{ck}}{1.5}, \frac{f_{sk}}{1.15}\right) \geqslant \begin{cases} 1.5\, S_{gk} + 1.5\, S_{qk} & (4.7) \\ 0.9\, S_{gk} + 1.5\, S_{qk} & (4.8) \\ 1.5\, S_{gk} + 1.35\, S_{q1k} + 1.20\, S_{q2k} \\ \quad + 1.05\, (S_{q3k} + S_{q4k} + \dots) & (4.9) \end{cases}$$

where:

S_{gk} represents the load effect due to the characteristic permanent loads,

S_{qk} represents the load effect due to the most critical characteristic variable load,

S_{qik} denote the load effects due to the ith characteristic variable load (with $S_{q1k} > S_{q2k} > S_{q3k} > \dots$)

Equation (4.8) must be used as an alternative to (4.7) when the contribution of the permanent loads is favourable to safety. Other terms must be added in (4.7), (4.8), and (4.9) in the presence of imposed deformations [48 - R. 22, 212] and suitable checks must be provided for structures subject to accidental actions [48 - R. 22, 211].

For service limit states, (4.7), (4.8), and (4.9) become:

$$g_R(f_{ck}, f_{sk}) \geq \begin{cases} S_{gk} + S_{qk} & (4.10) \\ S_{gk} + 0.9\,S_{q1k} + 0.8\,S_{q2k} \\ \quad + 0.7(S_{q3k} + S_{q4k} + \ldots) & (4.11) \end{cases}$$

In the presence of functional relations:

$$g_S(F_1, F_2, \ldots, F_n) \tag{4.12}$$

that are not proportional from the analytical viewpoint, (4.6) provides for a hookean linearisation corresponding to the values $\gamma_{f1i}\,\gamma_{f2i}\,F_{ik}$ of the actions. In fact the optimal intensity of the actions at which this linarisation should take place depends upon the respective mutual variabilities of the resistances and action effects [56]. Therefore it might be convenient to separate the definition of γ_{f3} factors from that of the intensities of the actions at which linearisation is effected. To this purpose it was suggested [57, 58] that such operations should be carried out corresponding to suitable values $a_i F_{ik}$, usually different from $\gamma_{f1i}\,\gamma_{f2i}\,F_{ik}$, and to replace then (4.12) by the proportional relation $g_{So}(\cdot)$ defined by:

$$\begin{cases} g_{So}(a_1 F_{1k}, a_2 F_{2k}, \ldots, a_n F_{nk}) \\ \quad = g_S(a_1 F_{1k}, a_2 F_{2k}, \ldots, a_n F_{nk}) \\ g_{So}(F_1=0, F_2=0, \ldots, F_n=0) = 0 \end{cases}$$

The γ_m and γ_f factors that appear in the safety checking relations examined above have been determined primarily on the basis of empirical considerations, that is, on the basis of experience gained from the observed behaviour of a wide variety of structures designed by classical calculation methods. However, apart from the problem of the combination of actions, it is possible, if only in special cases, to introduce a probabilistic interpretation for these coefficients. Always within the failure criterion (4.5), let:

$$r = g_R(f) , \tag{4.13}$$

be an increasing monotonic function of one material strength only, and:

$$s = g_S(F) = a\,F , \tag{4.14}$$

a proportional relation of one action F only. In this case, with reference to ultimate limit states, (4.6) becomes:

$$g_R(f_{0.05}/\gamma_m) - \gamma_{f3} g_S(\gamma_{f1} F_{0.95}) \geqslant 0 , \tag{4.15}$$

i.e.,

$$g_R(f_{0.05}/\gamma_m) - \gamma_f a F_{0.95} \geqslant 0 ; \tag{4.16}$$

if

$$\gamma_f = \gamma_{f3} \gamma_{f1}$$

and if the values 0.05 and 0.95 are the orders of the fractiles of the distributions of the random variables involved.

On the assumption, acceptable at times, that it may be possible to write:

$$f_{0.05}/\gamma_m = f_{0.005} , \tag{4.17}$$

taking into account (4.13) and (4.14), (4.16) becomes:

$$r_{0.005} - \gamma_f s_{0.95} \geqslant 0 . \tag{4.18}$$

Then, on the basis of (2.45):

$$\gamma_f = \gamma^* , \tag{4.19}$$

and from the curves of Fig. 2.7 (r and s both being normal), c_R being equal to ca. 0.15 and c_S ranging from 0 to 0.30, one can see that the probability of failure corresponding to the value $\gamma_f = 1.5$ established in the CEB-FIP Recommendations is $p_f \cong 10^{-5}$.

With the same positions as in the example above, but with reference to service limit states ($\gamma_m = 1$), (4.16) becomes:

$$r_{0.05} - \gamma_f s_{0.95} \geqslant 0 . \tag{4.20}$$

Then, on the basis of (2.44):

$$\gamma_f = \gamma_k , \tag{4.21}$$

and from the curves of Fig. 2.6 to $\gamma_f = 1$ (assumed in the CEB-FIP Recommendations) the corresponding value of p_f will range between 10^{-1} and 10^{-2} for any reasonable value of c_R and c_S.

Since these failure probabilities coincide with the target probabilies for the ultimate and service limit states respectively, it follows that, in this particular case and with the given restrictive assumptions, the foregoing scheme can be interpreted probabilistically. In reality, within the range of possible variations of c_R and c_S and for any distribution of R and S, for the value:

$$\gamma^* = 1.5$$

the corresponding limits of p_f are sensibly wider. Furthermore, in the general case, (4.17) is not logically satisfied. In any event, the semiprobabilistic limit states method proposed by the CEB-FIP Recommendations has undeniable advantages in comparison to traditional safety checking procedures based on the allowable stresses criterion, in spite of its theoretical limitations in terms of reliability, particularly since it takes into account all possible states of behaviour of the structures, considers the random character of the variables involved and permits a more satisfactory treatment of nonlinear problems. However, a number of factors have not been fully resolved, such as:

—the definition and determination of characteristic values,
—combination of actions varying with time,
—the values of partial safety factors,
—whether or not to use graduated coefficients.

These have given rise to great variations in the representation of identical physical phenomena in the codes adopted by different· countries drafted on the basis of the CEB-FIP 1970 Recommendations (for instance, variations from 20 to 40% in load effects [59]).

Indeed, remarkable progress has been made with respect to the 1970 Recommendations, especially with the publication of the manual on structural safety by H. Mathieu[59], and with the development of successive drafts of the "Common unified rules for different types of construction and material" [50, 10]. These documents and proposals have contributed to the discussion and putting into focus of various specific problems, especially those on the idealisation of time-varying actions, the definition of their characteristic values and the explicit consideration of the concept of the duration of the actions. In particular, leaving unchanged the definition of the characteristic values of the strengths of materials as fractiles of the order of 0.05 of their distributions, so as to be able to deal with the various situations that may be encountered, within the framework of the chosen method, the number of characteristic values for the actions had to be increased. Among these values, the most important are:

—the characteristic values F_k —
For nonstructural permanent loads, these are conceived as 0.95 or 0.05 fractiles of their respective statistical distributions; the former must be used when the effects are unfavourable with respect to the attainment of the limit state being considered, the latter being used for the opposite case. These two characteristic values G_{max}

and G_{min} can be replaced by the mean value if they do not depart by more than 5% from the latter. The self-weight of the structural elements is represented by a single nominal value, to be computed on the basis of the mean unit weight of the materials. The actions of prestress can be represented by two characteristic values, a maximum and a minimum value, which are considered as 0.85 and 0.15 fractiles. With regard to actions varying with time, for which the distributions of the maximum in 50 years have coefficients of variation not much greater than 0.20, the characteristic values F_k are characterised by a mean return period equal to ca. 120 years; if the coefficients of variation are higher than 0.20, the characteristic values F_k are so defined that the product 1.4 F_k should have a mean return period ranging between 1,000 and 10,000 years.

–combination values $\psi_0 F_k$ (appearing in fundamental safety verifications for ultimate limit states)–

For this determination use was made of the concept of the elementary interval and operational statistical methods expressed especially within a framework of linear functions of two or more random variables [59].

–"quasipermanent" values $\psi_2 F_k$ and "frequent" values $\psi_1 F_k$ (these are used for checking safety relationships concerning serviceability limit states and in the case of accidental combinations for ultimate limit states)–

The former correspond, as a general rule, to the mean values of the "as sampled" distributions of the instantaneous values of the actions (see 3.7) while the latter correspond to the 0.95 fractiles of the same distributions.

The chosen model, in a symbolic rather than algebraic representation, is of the following type [10]:

a) ultimate limit states (fundamental combinations)

$$g_R (f_{1k}/\gamma_{m1}, \ldots, f_{hk}/\gamma_{mh}) \geqslant \gamma_{f3} g_S [\, 1.2 \sum G_{max} + 0.9 \sum G_{min}$$

$$+ 1.4(F_{1k} + \sum_{i=2}^{n} \psi_{0i} F_{ik})]$$

where:

G_{max}, G_{min} denote the representative values of permanent actions, unfavourable and favourable with regard to the attainment of the limit state under consideration, respectively

F_1 denotes the basic action in the combination
$F_i (i=2,3,...,n)$ denote the complementary actions
$f_j (j=1,2,...,h)$ denote the material strengths.

Each of the variable actions in turn must be considered as the basic action (and the others as the complementary actions) with the aim of finding the most critical combination with regard to the limit state being considered. Further checks must also be carried out to take account of accidental actions [10, Section 10.3.1].

b) serviceability limit states:

$$g_R (f_{1k}, \ldots, f_{hk}) \geqslant \begin{cases} g_S [\bar{G} + F_{1k} + \displaystyle\sum_{i=2}^{n} \psi_{1i} F_{ik}] & (b_1) \\[2em] g_S [\bar{G} + \psi_1 F_{1k} + \displaystyle\sum_{i=2}^{n} \psi_{2i} F_{ik}] & (b_2) \\[2em] g_S [\bar{G} + \displaystyle\sum_{i=1}^{n} \psi_{2i} F_{ik}] & (b_3) \end{cases} \quad .$$

where $\bar{G}$ represents the mean value of the permanent actions. Expr. (b_1) represents the infrequent combinations, (b_2) the frequent combinations, (b_3) the quasi-permanent ones.

The values of the coefficients ψ_0, ψ_1, ψ_2 relating to the greater part of the variable actions encountered most frequently in structural problems are given in tables in CEB bulletin n° 116 E[10]. The coefficients γ_{mi} for the various materials and limit states, and the coefficient γ_{f3}, are at present in course of being defined, but suitable values for these coefficients have been proposed for concrete structures [60].

These new proposals represent significant progress compared with the CEB-FIP 1970 Recommendations. Nevertheless, for the general case it is not easy to deduce accurate information with these methods on the probability of failure characterising a given structural design. Even these safety factors, in fact, have been obtained more by "engineering judgement" than on the basis of the methods of reliability analysis. This is compatible however with the basic principles of level 1. An alternative approach for the determination of the partial safety factors would be to derive them from level 2 methods.

4.3 DERIVATION OF PARTIAL SAFETY FACTORS FROM LEVEL 2 METHODS

With reference to the concepts outlined in Part 3, let:

$$\underline{X}[X_1, X_2, \ldots, X_n]$$

be the vector of the random variables involved in a general structural problem and let:

$$g(x_1, x_2, \ldots, x_n) = 0 \qquad (4.22)$$

represent the safety domain boundary.

Introducing the standardised variables:

$$u_i = \frac{x_i - \bar{x}_i}{\sigma_{X_i}} \ , \ (i = 1, 2, \ldots, n) \ , \qquad (4.23)$$

into the new reference, (4.22) becomes:

$$g(u_1, u_2, \ldots, u_n) = 0 \ . \qquad (4.24)$$

The point defined by the coordinates: $u_1 = 0, u_2 = 0, \ldots, u_n = 0$ belongs to the safety domain boundary, that is:

$$g(u_1 = 0, u_2 = 0, \ldots, u_n = 0) > 0 \ .$$

Moreover, let the variables X_i have multinormal joint density probability and be stochastically independent. Such assumptions are not as narrow as they seem at first sight, since:

—if there were interdependence among the X_i's which could be expressed in terms of a known covariance matrix, it would always be possible to replace them by uncorrelated ones by means of an orthogonal transformation,

—if the X_i variables had general non-normal distributions, it would be possible to replace them by gaussian variables through suitable exact or approximate transformations.

Now, in accordance with the Lind-Hasofer method, the reliability index β is defined as the minimum distance $0P^*$ in (4.24) from the origin 0. Fig. 4.1 illustrates a two-dimensional case.

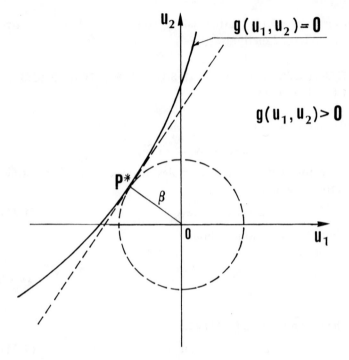

Fig. 4.1

Within the approximations indicated in 3.4, it is possible to associate the "operational" probability p_f of attaining the limit state with the minimum distance β by means of the relationship:

$$p_f = \Phi(-\beta) . \tag{4.25}$$

This positition is equivalent to the replacement of (4.24) by the hyperplane passing through P* and normal to 0P*. Indicating by

$$\underline{a}[a_1, a_2, \ldots, a_n]$$

the direction cosines of the straight line 0P* oriented toward the failure region, the coordinates of the "design point" P* will be:

$$u_i^* = a_i \beta \quad (i = 1, 2, \ldots, n)$$

thus satisfying the relationship:

$$g(u_1^*, u_2^*, \ldots, u_n^*) = g(a_1\beta, a_2\beta, \ldots, a_n\beta) = 0 \tag{4.26}$$

The coordinates of P*, projected in the space of the X_i variables, will be:

$$x_i^* = \bar{x}_i + a_i \beta \sigma_{X_i} = \bar{x}_i (1 + a_i \beta c_{X_i}) \ , \tag{4.27}$$

and (4.26) becomes:

$$g(x_1^*, x_2^*, \ldots, x_n^*) =$$

$$g[\bar{x}_1(1 + a_1 \beta c_{X_1}), \bar{x}_2(1 + a_2 \beta c_{X_2}), \ldots, \bar{x}_n(1 + a_n \beta c_{X_n})] = 0 \tag{4.28}$$

Working with level 1 methods, with reference to (4.2), the design equation can be written in the form:

$$g(\Delta_1 \bar{x}_1, \Delta_2 \bar{x}_2, \ldots, \Delta_n \bar{x}_n) = 0 \ , \tag{4.29}$$

and

$$\Delta_r = \gamma_{mr}^{-1} (1 - k_r c_{X_r}) \quad (r = 1, 2, \ldots, h) \tag{4.30}$$

$$\Delta_s = \gamma_{fs}(1 + k_s c_{X_s}) \quad (s = h+1, h+2, \ldots, n) \tag{4.31}$$

Equation (4.29) shows, among other things, that it is possible to introduce partial safety factors related to the mean values instead of to characteristic values. A comparison between (4.29) and (4.28) shows that the same design, calculated with level 2 and level 1 methods, is characterised by the same degree of reliability if the relations:

$$\Delta_i = 1 + a_i \beta c_{X_i} \ , \qquad (i = 1, 2, \ldots, n) \tag{4.32}$$

that is, for (4.30) and (4.31):

$$\gamma_{mr}^{-1} = (1 + a_r \beta c_{X_r})/(1 - k_r c_{X_r}) \ , \quad (r = 1, 2, \ldots, h) \tag{4.33}$$

$$\gamma_{fs} = (1 + a_s \beta c_{X_s})/(1 + k_s c_{X_s}) \ , \quad (s = h+1, h+2, \ldots, n) \tag{4.34}$$

are satisfied.

Therefore, with reference to a fixed limit state function, level 1 methods can be made identical to those of level 2, if the γ_m and γ_f factors are expressed by means of the functional relationships (4.33) and (4.34). This operation can be referred to as "derivation of the partial safety factors on the basis of the design point coordinates". Direct use of (4.33) and (4.34) however is not advisable in level 1 code provisions which necessarily call for the use of constant γ_m and γ_f factors for a wide range of design situations. From the practical standpoint, it is therefore necessary to discretise these continuous functions appropriately. To better illustrate this problem it is useful to examine first some aspects of the different possibilities of deriving partial safety factors within

linear limit state equations [61].
Thus, let:

$$\sum_{i=1}^{n} a_i x_i = 0 \qquad (4.35)$$

be the safety domain boundary, expressed in the space of the basic random variables X_i.

A common design situation consists of determining a parameter of one of the random variables, for example $\bar{x}_1$, knowing all the other statistical parameters, as well as the deterministic quantities involved. Let us determine this design parameter while operating at level 1 on the basis of a given set of coefficients Δ_i, that is by solving the equation:

$$\sum_{i=1}^{n} a_i \Delta_i \bar{x}_i = 0 \ . \qquad (4.36)$$

with respect to $\bar{x}_1$.

In the space of the standardised variables U_i, the design under consideration is characterised by the hyperplane of the equation:

$$\sum_{i=1}^{n} a_i \sigma_{X_i} u_i + \sum_{i=1}^{n} a_i \bar{x}_i = 0 \ . \qquad (4.37)$$

Let $\beta = OP^*$ be the reliability index and:

$$\underline{a}[a_1, a_2, \ldots, a_n]$$

the versor of the straight line OP^*.

Fig. 4.2 shows the two-dimensional case. It is:

$$\underline{OP}^* = \underline{a}\beta \qquad (4.38)$$

Now let P be any point of the hyperplane (4.37) and

$$\underline{\delta}[\delta_1, \delta_2, \ldots, \delta_n]$$

a suitable vector (not a unit vector, excepting the case in which $P \equiv P^*$), such that:

$$\underline{OP} = \underline{\delta}\beta \qquad (4.39)$$

Projecting $\underline{OP}$ on the direction OP^*, it follows that:

$$\sum_{i=1}^{n} a_i \delta_i = 1 \ . \tag{4.40}$$

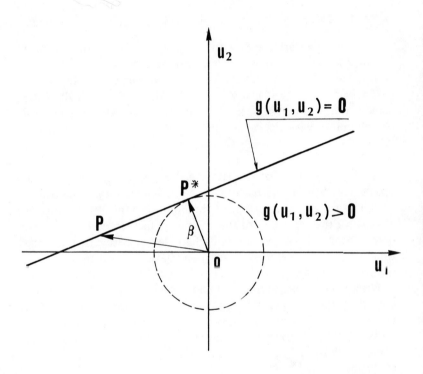

Fig. 4.2

The image of P in the space of the variables X_i will have the coordinates:

$$\bar{x}_i + \delta_i \sigma_{X_i} \beta \ , \quad (i = 1, 2, \ldots, n) \tag{4.41}$$

which satisfy equation (4.35). Therefore:

$$\sum_{i=1}^{n} a_i \bar{x}_i (1 + \beta \delta_i c_{X_i}) = 0 \tag{4.42}$$

Comparing (4.42) with (4.36) one obtains:

$$\Delta_i = 1 + \beta \delta_i c_{X_i} \ , \quad (i = 1, 2, \ldots, n) \tag{4.43}$$

and, on the basis of (4.30), (4.31):

$$\gamma_{mr}^{-1} = (1 + \beta \delta_r c_{X_r})/(1 - k_r c_{X_r}) \ , \quad (r = 1, 2, \ldots, h) \tag{4.44}$$

$$\gamma_{fs} = (1 + \beta \delta_s c_{X_s})/(1 + k_s c_{X_s}) \ , \quad (s = h+1, h+2, \ldots, n) \ . \tag{4.45}$$

Point P being arbitrary, with the only condition that it must belong to (4.37), there exist an infinite number of $\underline{\delta}$ vectors, the components of which satisfy (4.40), and hence infinite sets of partial factors Δ_i (4.43) or γ_{mr}, γ_{fs} (4.44), (4.45) which allow the elaboration of a design at level 1 with the same reliability index β and therefore with the same reliability $1 - \Phi(-\beta)$. In other words all the $\underline{\delta}$ vectors satisfying (4.40) give the same design as the $\underline{a}$ versor.

In the framework of similar considerations, a number of proposals have been made recently [62, 63] which suggest the choice of points different from that of minimum distance P*, to serve as a basis for the derivation of the γ_m and γ_f factors, with the aim of determining sets of partial coefficients less sensitive to the variations of the parameters involved than those defined by means of (4.33) and (4.34).

Within the infinite number of choices of the Δ_i coefficients that give the same design, it is logically possible, from a theoretical viewpoint, to fix some in advance at a trivial value, for instance unity. For $\Delta_i = 1$, expr. (4.43) gives $\delta_i = 0$. Assuming p of the n Δ_i coefficients to have a value equal to unity and setting $q = n - p$, (4.40) becomes:

$$\sum_{i=1}^{q} a_i \delta_i = 1 \ . \tag{4.46}$$

Therefore a level 1 code may require unit safety factors (related to mean values) for a predetermined number of random variables, such as for instance: geometric parameters, imposed deformations, permanent loads.

Analogous considerations are also valid for nonlinear limit state functions, in the hypothesis of replacing the safety domain boundary in the space of the standardised variables by the hyperplane tangent at P* to the hypersphere centred at the origin, and referring to the operational failure probability (4.25).

4.4 PRACTICAL DETERMINATION OF THE PARTIAL SAFETY FACTORS

As seen in 4.3, level 1 methods can be made identical in terms of reliability to those of level 2 if the partial safety factors are expressed as continuous functions of the statistical parameters of the basic variables on the basis of (4.32), (4.33) and (4.34). The factors determined in this manner cannot, however, be used conveniently within a level 1 code, since these methods require that constant coefficients should be available for a wide range of design situations. It is therefore necessary to search for suitable methods of discretisation. Moreover, for practical reasons, a number of restrictions condition the choice of safety factors. Among these can be listed:

—partial safety factors should not be less than unity

—partial coefficients for the actions should be the same for structures built with different materials and of different types of construction

—to simplify the operational procedure some factors can be taken with unit values; values greater than 1 should however be associated with all the random variables X_i characterised by large values of the sensitivity coefficients a_i.

Such a discretisation of the coefficients can be carried out in various ways, even within the framework of the above restrictions. Thus, it is possible to use simple operational procedures which do not require the explicit calculation of the probability of attaining the limit state, as for level 2 and for level 3; the penalty that must be paid is that the operational reliability of the basic design obtained in this manner will then deviate to a greater or lesser extent from the target level of the code. Therefore, the most important problem is that of choosing sets of partial safety factors in such a way that the deviations from the target reliability levels will be minimal and, if possible, contained within sufficiently narrow prescribed limits.

As stated in recent work by Lind [53], a practical procedure for determining appropriate sets of partial safety factors for use in codes involves taking into account the following concepts and operations:

—definition of the scope of the code and also that of suitable "subcodes", conceived as a parametered set of the structures or structural elements governed by them. The various design situations which are regulated by a code can be characterized by attributes (such as type of structure, materials of which it consists, specific limit state functions, etc.) and parameters such as beam spans, depth-breadth ratios, the ratios of permanent loads to variable loads, percentage of compression reinforcement, etc.). A subcode applies to a well defined set of design situations covered by the code. For operational reasons the range of variation of the parameters concerned is subdivided into classes of appropriate size. For instance, with regard to the span of reinforced concrete beams, the following classes might be envisaged: 3 to 5 m, 5 to 7 m, . . . , 13 to 15 m; each class is then denoted by the value corresponding to the middle value of the interval concerned. Thus a subcode is represented by a finite number of specific design conditions; let m be this number. For each subcode a certain group of partial safety factors must be determined. The number m of design situations envisaged can be either large or small; if m is large, it will be necessary to make use of a complex level 1 model and expect designs having a measure of reliability appreciably divergent from the target level, which is understood to be the objective to be attained by the code; if it is small, the format can be simpler and satisfy more restricted requirements of homogeneity in terms of reliability. The selection of the scope, and hence, of the number of subcodes to be considered, requires a balance between the desired operational simplicity and the close attainment of the objective of the code.

−choice of the reliability levels to serve as the operational target
−estimate of the expected relative frequencies w_i of each of the m design situations foreseen by the subcode. Obviously,

$$\sum_{i=1}^{m} w_i = 1 \ .$$

Since the target reliability cannot be achieved exactly for the various design situations, it is necessary to know which of these will be most frequently encountered, so that greater importance can be attributed to these in assessing how closely the objective of the code has been attained. These estimates can be effected either by extrapolation from observations of structures constructed in the past, or on the basis of reasoned judgement.

−choice of the degree of efficiency of the subcode, i.e., definition of the allowable deviations from the target reliability. A suitable procedure might be to require that the differences between target probability of failure and the operational failure probabilities for each design based on a fixed group of partial safety factors, should not exceed given quantities, for instance $\pm\ 10^{-1}$. An alternative procedure would consist of fixing an upper bound on the sum S of the squares of the differences between the logarithms of the operational probabilities of a design $(p_f)_i$ and the logarithm of the target probability $p_f{}^*$, weighting each of these on the basis of its frequency of occurrence w_i. A limit value of S now being proposed [10] is:

$$S = \sum_{i=1}^{m} w_i (\log p_f - \log p_f{}^*)_i^2 \leqslant 0.25 \ . \qquad (4.47)$$

Operational procedures for the determination of partial safety factors founded on these concepts have been recently developed in the United Kingdom, in connection with a study of probabilistic limit states structural codes [54]. The basic operational steps of this method, with special reference to structures subjected to permanent loads and only one load varying with time, are the following:

a) definition of the scope of the subcodes
b) subdivision of the respective parameters into classes of approximately constant size and determination of the number m of design situations to be considered
c) estimate of the expected relative frequencies w_i to be assigned to each of the m design situations

d) using an existing code design each of the m structural elements, referring to the values of the parameters corresponding to the middle value of each of the respective classes concerned
e) using a suitable level 2 method, determine for each structural element, dimensioned according to d), the operational failure probability p_f and furthermore establish, again for each element, the relationship between p_f and one of the most important design parameters ϑ (for reinforced concrete beams, ϑ could be the percentage of tension reinforcement) within a reasonable range of failure probabilities (for instance, $10^{-3} < p_f < 10^{-9}$). The theoretical functional relationship assumed in [54] is:

$$\log p_f = a_0 + a_1 \vartheta + a_2 \vartheta^2 \qquad (4.48)$$

in which the constants a_0, a_1, a_2 must be determined for each design.

f) choose the desired level 1 format, select which of the random variables must be represented by their mean values (or nominal values, in a broad sense), define the characteristic values of the remaining p variables

g) choose suitable initial values for the p partial safety factors

h) solve a design problem at level 1, that is, determine the value of the design parameter ϑ, for each of the m structural elements, with the values of the partial safety factors obtained from g)

i) calculate by means of (4.48) the value of the operational failure probability p_f that corresponds to the ϑ value determined in h), for each of the m structural elements

j) compute S, the sum of the squares of the weighted differences on the basis of (4.47), and check that the resulting value is lower than the previously stated limit value. If this is not satisfied, then modify the initial values of the p partial safety factors using a suitable algorithm and repeat steps h) to j), until (4.47) is satisfied. If it is impossible to reach the limit value prescribed for S, then it will be necessary to resort to a more complex level 1 format (generally with a greater number of safety factors), or to reduce the number of the m design situations covered by the subcode.

Clearly, this procedure does not deal with the problem of the determination of the optimal level 1 format, but it solves that of the determination of rational and, from an engineering point of view, acceptable sets of partial safety factors.

For structures subject to more than one action which vary with time, the current level 2 operational methods take into account the k different potentially critical combinations identified in 3.7, that is, as many combinations as are actions involved. In this case, the operational failure probability is calculated on the basis of (3.94). By the hypothesis of adopting the same number of checks in level 1 methods, it is possible to derive, on the basis of this procedure, k different sets of partial safety factors. The calculation becomes more complicated than in the case of only one variable action's being present; it is nevertheless entirely analogous from the operational viewpoint. No actual specific experience exists with this method but it seems likely that for the safety analysis of most structures the designer would not be required to carry out all these checks, since some of the k combinations will be seen by inspection not to be critical for the procedures relating to the determination of the partial safety factors. For such a case, the code will have to specify the number and type of the combinations of loads to be considered.

Background to the Statistics and Probability Calculations

5.1 PREMISE

A number of terms, symbols or statistical concepts that appear in the previous chapters may not be familiar to the general reader. In Part 5, therefore, a number of definitions, results and operational procedures of statistics and probability calculation are recalled, omitting, in most cases, the detailed proofs for which the reader is referred to the many textbooks extant on the subject. Among these, the texts listed in the "References" under numbers [18], [64], and [65] are especially recommended, in connection with the problems dealt with in this study.

5.2 PROBABILITY OF RANDOM EVENTS

Let us agree, in a broad sense, to apply the term "events" to the results of experiments or observations effected under defined conditions; where these events cannot be predicted with certainty, they are referred to as "random events". The "relative frequency" of a given event is the ratio between the number of tests in the course of which that event is found to have occurred and the total number of tests.

Experience teaches that, as the number of tests increases, the relative frequency of a random event will tend to become stabilized near a certain value. The "probability of an event" can be conceived as the idealisation of the concept of its relative frequency, i.e., according to Von Mises [66], it is the limit the relative frequency tends to reach, as the number of tests tends to infinity. This definition of probability, among the many that have been proposed, is

perhaps best suited to the requirements of the application of probability concepts to the experimental sciences; however it is not
wholly free from criticisms: one of these is that it is not always
possible to make the number of tests tend to infinity. It is certainly
beyond the aims of these background notes to develop this problem; let it only be noted here that it is not essential to define the
concept of probability in order to carry out probability calculations, just as it is not necesary to define point, straight line, and
plane explicitly to solve problems in terms of Euclidean geometry.
It is, however, essential to establish a set of rules to be respected
whenever the concept of probability is introduced, whatever its
precise definition may be. In this way was born the axiomatic
theory of probability, whose main exponent is Kolmogorov [67].
In its context, let S be the "space of elementary events", the
structure of which depends on the specific problem being
examined; it can be discrete, i.e., contain a finite (or denumerably
infinite) number of points, or continuous. A "Borel set" is defined
as the class B of the subsets characterised by the following properties:

1) the space S belongs to B,
2) the empty set $\emptyset$ belongs to B,
3) if a finite (or denumerably infinite) number of subsets of S belongs to B, their union and their intersection also belong to B,
4) if two subsets of S belong to B, their difference also belongs
to B.

More rigorously, an event is defined as every subset of S belonging to a class having properties 1, 2, 3, and 4.

Given a space S and a set of random events associated with S,
the following axioms are universally accepted:

a) The probability of event A, denoted by $P\{A\}$, is a real number
such that:

$$0 \leqslant P\{A\} \leqslant 1 \ . \tag{5.1}$$

b) The probability of an event that is certain to occur (corresponding to the entire space S) is 1, that is:

$$P\{S\} = 1 \tag{5.2}$$

c) The probability of the union of a finite number or of a denumer-
able infinity of disjointed subsets of S (corresponding, that is to say,
to mutually exclusive events) is equal to the sum of the probabilities
of these events:

$$P\{A_1 + A_2 + A_3 + \ldots\} = P\{A_1\} + P\{A_2\} + P\{A_3\} + \ldots \quad (5.3)$$

Axiom (5.3) is also currently referred to as the addition rule.

Using axioms (5.1), (5.2) and (5.3) and some results of set theory, it is easy to prove the following theorems:

−The probability of an impossible event (corresponding to the empty set $\emptyset$), is equal to 0:

$$P\{\emptyset\} = 0 \qquad\qquad (5.4)$$

−The probability of an event A is the complement to 1 of that of the contrary event $\bar{A}$ (corresponding to the set complementary to A):

$$P\{A\} = 1 - P\{\bar{A}\} \ . \qquad\qquad (5.5)$$

−The probability of the union of two events is equal to the sum of the individual probabilities minus that of their intersection:

$$P\{A + B\} = P\{A\} + P\{B\} - P\{AB\} \ , \qquad\qquad (5.6)$$

which is the theorem of total probability. If A and B are mutually exclusive, then $AB = \emptyset$ and (5.6) is reduced to a particular formulation of axiom (5.3).

5.2.1 Conditional Probability and Stochastic Independence

Let the symbol $P\{A/B\}$ denote the conditional probability of event A, given that the event B had occurred, i.e. the probability of the occurrence of event A on the assumption that event B has already taken place, then (see for example [68]):

$$P\{A/B\} = P\{AB\}/P\{B\} \ , \qquad\qquad (5.7)$$

and by analogy:

$$P\{B/A\} = P\{AB\}/P\{A\} \ . \qquad\qquad (5.8)$$

From (5.7) and (5.8) it follows that:

$$P\{AB\} = P\{A\}P\{B/A\} = P\{B\}P\{A/B\} \ , \qquad\qquad (5.9)$$

a relation known as the joint probability theorem.

Two events are defined as "stochastically independent" when the occurrence or non-occurrence of one does not affect the probability of the other, that is, when:

$$P\{A/B\} = P\{A\} \ , \qquad\qquad (5.10)$$

as well as:

$$P\{B/A\} = P\{B\} \ . \tag{5.11}$$

Hence, in the case of stochastically independent events, (5.9) becomes:

$$P\{AB\} = P\{A\}P\{B\} \ , \tag{5.12}$$

i.e. the joint probability $P\{AB\}$ of two indpendent events A and B is equal to the product of the probabilities associated with the individual events.

In the case of n stochastically independent events (5.12) can be written in generalised form:

$$P\{A_1 \ A_2 \dots A_n\} = P\{A_1\} \ P\{A_2\} \dots P\{A_n\} \tag{5.13}$$

and it is then called the product rule.

5.2.2 Total Probability Theorem and Bayes's Theorem

Let the space S of the elementary events be divided into n sub-sets corresponding to the mutually exclusive events $B_1, B_2, \dots, B_n$ (their union is, by definition, equal to S) and consider a random event A.

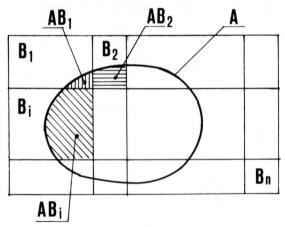

Fig. 5.1

Then (see Fig. 5.1):

$$A = A\,B_1 + A\,B_2 + \dots + A\,B_n \ , \tag{5.14}$$

$$P\{A\} = P\{A\,B_1 + A\,B_2 + \dots + A\,B_n\} \ ,$$

and therefore, since the sets A B_i are disjoint, on the basis of (5.3):

$$P\{A\} = \sum_{i=1}^{n} P\{A\,B_i\} \tag{5.15}$$

Applying (5.9) to each term of the second member of (5.15) it follows that:

$$P\{A\} = \sum_{i=1}^{n} P\{A/B_i\}P\{B_i\} \tag{5.16}$$

This result expresses the absolute probability theorem.

Continuing the study of event A and of the mutually exclusive and collectively exhaustive events $B_1, B_2, \ldots, B_n$, on the basis of (5.9) one gets:

$$P\{B_i/A\} = \frac{P\{A/B_i\}\,P\{B_i\}}{P\{A\}} \tag{5.17}$$

Substituting (5.16) into (5.17):

$$P\{B_1/A\} = \frac{P\{A/B_i\}\,P\{B_i\}}{\displaystyle\sum_{i=1}^{n} P\{A/B_i\}\,P\{B_i\}} \tag{5.18}$$

Expression (5.18) is also called Bayes's rule and is frequently used in engineering applications. If the "a priori probabilities" $P\{B_i\}$ of the various possible causes B_i and the conditional probabilities $P\{A/B_i\}$ associated with the various possible outcomes are known, Bayes's theorem permits to state, "a posteriori", which, among the B_i causes, is the most likely one: namely the one that maximises $P\,B_i/A$ (Bayes's estimate).

5.3 ONE—DIMENSIONAL RANDOM VARIABLES

If every simple event of a set S is made to correspond to a real number x, then this defines a random variable X. As a general rule random variables are denoted by capital letters and their argumental values by small letters. The properties of a random variable may be characterised by its cumulative distribution function $F(x)$ defined by:

$$F(x) = P\{X \leqslant x\} \quad . \tag{5.19}$$

Random variables may be discrete and continuous.

Discrete random variables are those that can assume x_i values belonging to a discrete finite or denumerably infinite set (i.e. such as can be put in a biunivocal correspondence with the set of natural numbers).

$$p(x_i) = P\{X = x_i\} \tag{5.20}$$

is then defined as the probability mass function.

If the number of the argumental values: $x_1, x_2, \ldots, x_n$ is finite, then:

$$\sum_{i=1}^{n} p(x_i) = 1 \tag{5.21}$$

If, on the other hand, the x_i's constitute a denumerably infinite set, then:

$$\sum_{i=1}^{\infty} p(x_i) = 1 \quad .$$

According to (5.19), the cumulative distribution function is given by:

$$F(x) = \sum_{x_i \leqslant x} p(x_i) \quad , \tag{5.22}$$

and it is a stepped function, monotone, nondecreasing and lying between the values 0 and 1.

Continuous random variables are those that assume x argumental values belonging to a continuous set, whose limits may be even $-\infty$ and $+\infty$. Their cumulative distribution $F(x)$ is continuous, and there exists a function $f(x)$, called probability density, such that:

$$F(x) = \int_{-\infty}^{x} f(z) \, dz \quad . \tag{5.23}$$

At the points where $F(x)$ is derivable:

$$f(x) = \frac{d\,F(x)}{dx} \tag{5.24}$$

The following relationships apply:

$$f(x) \geqslant 0 \tag{5.25}$$

$$P\{x < X \leqslant x+dx\} = f(x)\,dx \tag{5.26}$$

$$\int_{x_1}^{x_2} f(z)\,dz = F(x_2) - F(x_1) \tag{5.27}$$

$$F(-\infty) = 0 \tag{5.28}$$

$$F(+\infty) = 1 \tag{5.29}$$

$$\int_{X}^{+\infty} f(z)\,dz = 1 - F(x) \tag{5.30}$$

5.3.1 The Indices of Location and of Scatter for a Random Variable

The expected value (or mean value, or simply the mean) of a continuous random variable X is defined by the integral:

$$E(x) = \int_{-\infty}^{+\infty} x\,f(x)\,dx \tag{5.31}$$

which is often denoted also by the symbol $\bar{x}$.

The expected value of a function $\varphi(x)$ of the variable X is :

$$E[\varphi(x)] = \int_{-\infty}^{+\infty} \varphi(x)\,f(x)\,dx \ .$$

For discrete variables (5.31) becomes:

$$E(x) = \sum_{i=1}^{n} x_i\,p(x_i) \ . \tag{5.32}$$

The a fractile ($0 \leqslant a \leqslant 1$) of the distribution of the continuous random variable X is the number x_a solution of:

$$F(x) = a \qquad (5.33)$$

In particular, the median $\check{x}$ is the fractile of order $1/2$. Thus:

$$F(\check{x}) = 1/2 \ . \qquad (5.34)$$

The mode $\tilde{x}$ of a unimodal distribution is the value of X possessing maximum probability density:

$$f(\tilde{x}) = \max \ . \qquad (5.35)$$

If there is more than one relative maximum, each abscissa of these points is a mode and the distribution is called multimodal.

Using the concept of mean value, it is possible to define entire classes of parameters of particular interest. Thus, the mean of the kth power of X is defined as simple K order moment:

$$m_k = E(x^k) = \int_{-\infty}^{+\infty} x^k \, f(x) \, dx \ . \qquad (5.36)$$

In particular:

$$m_1 = \bar{x} \ . \qquad (5.37)$$

The central moments of order k are defined by the relationship:

$$\mu_k = E[(x - \bar{x})^k] = \int_{-\infty}^{+\infty} (x - \bar{x})^k \, f(x) \, dx \ . \qquad (5.38)$$

The first central moment is zero; that of the second order is called variance and usually denoted by the symbol σ^2:

$$\sigma^2 = \mu_2 = \int_{-\infty}^{+\infty} (x - \bar{x})^2 f(x) \, dx \ . \qquad (5.39)$$

The symmetry of a distribution is given by the fact that the central moments of odd order are zero. To measure the deviation of a distribution from the symmetrical, one can refer to the coefficient of skewness:

$$\gamma = \mu_3 / (\mu_2)^{3/2} \ ; \qquad (5.40)$$

depending on the sign of γ one can refer to positive or negative skewness.

The positive square root of the variance:

$$\sigma = \sqrt{\sigma^2} \tag{5.41}$$

is called standard deviation.

The coefficient of variation is the ratio:

$$c = \sigma/\bar{x} \ . \tag{5.42}$$

When considering a set of argumental values $x_1, x_2, \ldots, x_n$, which can always be interpreted as a statistical variable with frequencies $p(x_i)$ all equal to $1/n$, the expressions (5.31) and (5.39) become:

$$\bar{x} = \frac{1}{n} \sum_{i=1}^{n} x_i \ , \tag{5.43}$$

$$\sigma^2 = \sum_{i=1}^{n} (x_i - \bar{x})^2 / n \ . \tag{5.44}$$

With reference to a sample of n observations, the correct estimate of the population variance is given by:

$$\sigma_c^2 = \sum_{i=1}^{n} (x_i - \bar{x})^2 / (n - 1) \ .$$

Among the functions of a variable X, the *standardised variable U* is of particular interest; it is connected to the former by the linear relation:

$$u = (x - \bar{x})/\sigma \tag{5.45}$$

Hence the standardized variable is nondimensional, its mean is zero and its variance is equal to 1.

It should be noted that the mean value $\bar{x}$ of a random variable X yields a positional indication of the distribution, whereas the standard deviation is a parameter of the scatter, in the sense that knowledge of it permits us to deduce the degree of the distribution, i.e. whether it is more or less concentrated near the mean value. In this connection, the inequality of Tchebychef is of interest: without knowing anything about the distribution of a variable X except the mean $\bar{x}$ and the standard deviation σ, after fixing a positive real number λ, it follows that:

$$P\{\bar{x} - \lambda\sigma \leqslant X \leqslant \bar{x} + \lambda\sigma\} \geqslant 1 - 1/\lambda^2 \ , \qquad (5.46)$$

or, for $\epsilon = \lambda\sigma$:

$$P\{\bar{x} - \epsilon \leqslant X \leqslant \bar{x} + \epsilon\} \geqslant 1 - \sigma^2/\epsilon^2 \ . \qquad (5.47)$$

In the class of all the variables whose mean and variance alone are known, there is no inequality better than Tchebychef's[68]; this statement does not hold if assumptions are made on the shape of the distribution of the variable which is under consideration [69].

5.3.2 Theoretical Models for Continuous Random Variables

In this Section discussion is limited to the distribution functions and the related fundamental statistical parameters of the continuous random variables mentioned in Parts 1 to 4 of the text.

5.3.2.1 Normal distribution

—cumulative distribution function:

$$F(x) = \frac{1}{\sqrt{2\pi}\ \sigma} \int_{-\infty}^{x} e^{-(z-\bar{x})^2/2\sigma^2}\,dz$$

—probability density:

$$f(x) = \frac{1}{\sqrt{2\pi}\ \sigma}\ e^{-(x-\bar{x})^2/2\sigma^2}$$

—mean $\bar{x}$
—median $\check{x} = \bar{x}$
—mode $\tilde{x} = \bar{x}$
—standard deviation σ
  Where:

$$u = (x - \bar{x})/\sigma \ ,$$

then:

$$\Phi(u) = F_u(u) = \frac{1}{\sqrt{2\pi}} \int_{-\infty}^{u} e^{-t^2/2}\,dt$$

$$\varphi(u) = f_u(u) = \frac{1}{\sqrt{2\pi}}\ e^{-u^2/2}$$

It follows that:

$$\Phi(-u) = 1 - \Phi(u)$$

5.3.2.2 Log-normal distribution

−probability density:

$$f(x) = \frac{1}{x\sqrt{2\pi}\ \sigma_{\ln x}}\ e^{-[\ln(x/\check{x})]^2 / 2\sigma_{\ln x}^2}$$

−mean $\bar{x} = \check{x}e^{\sigma_{\ln x}^2/2}$
−median $\check{x}$

−mode $\tilde{x} = \check{x}e^{-\sigma_{\ln x}^2}$

−standard deviation $\sigma = \bar{x}[e^{\sigma_{\ln x}^2} - 1]^{1/2}$

5.3.2.3 Extreme type I distribution (maximum values)

−cumulative distribution function:

$$F(x) = e^{-e^{-a(x-\tilde{x})}}$$

−probability density:

$$f(x) = a\,e^{-a(x-\tilde{x})-e^{-a(x-\tilde{x})}}$$

−mean $\bar{x} = \tilde{x} + 0.5772/a$
−median $\check{x} = \tilde{x} - \ln(\ln 2)/a$
−mode $\tilde{x}$
−standard deviation $\sigma = \pi/\sqrt{6}\ a = 1.2825/a$
0.5772 is Euler's constant.
  Introducing the reduced variable $y = a(x - \tilde{x})$, one gets:

$$F_Y(y) = e^{-e^{-y}}$$

5.3.2.4 Extreme type II distribution (maximum values)

−cumulative distribution function:

$$F(x) = e^{-(kx)^{-\beta}} \quad (x \geqslant 0, \beta > 0, k > 0)$$

−probability density:

$$f(x) = \beta k (kx)^{-(\beta + 1)} e^{[-(kx)^{-\beta}]}$$

−mean $\bar{x} = \dfrac{1}{k} \Gamma(1 - 1/\beta)$

−median $\check{x} = \dfrac{1}{k} (-\ln \dfrac{1}{2})^{-(1/\beta)}$

−mode $\widetilde{x} = \dfrac{1}{k} \left[\dfrac{\beta}{\beta + 1} \right]^{1/\beta}$

−standard deviation:

$$\sigma = \frac{1}{k} [\Gamma(1 - 2/\beta) - \Gamma^2(1 - 1/\beta)]^{1/2}$$

$\Gamma(\cdot)$ is the Gamma function.

5.3.2.5 Chi-square function

This is the distribution of a continuous random variable, sum of the squares of ν normal, standardised, and independent variables. Its cumulative distribution function has been denoted in Part 3 by the symbol $\chi^2_\nu (\cdot)$.

−probability density:

$$f(x) = \frac{1}{2^{\nu/2} (\nu/2)} e^{-x/2} x^{(\nu/2) - 1}$$

−mean $\bar{x} = \nu$
−mode $\widetilde{x} = \nu - 2$
−standard deviation $\sigma = \sqrt{2\nu}$

5.3.3 Parameters of the Distribution of the Maximum Values

In load combination problems it is necessary to consider the distribution of the maximum of r random variables $X_1, X_2, \ldots, X_r$

independent and identically distributed with a common cumulative distribution function F(x). Denoting by $F_r(x)$ the cumulative distribution function of the maximum values, one obtains:

$$F_r(x) = [F(x)]^r$$

Let $\bar{x}$, σ and $\bar{x}_r$, σ_r be the means and standard deviations of the two distributions $F(x)$ and $F_r(x)$ respectively. The following cases can be distinguished:

a) If $F(x)$ is normal, then $F_r(x)$ is no longer normal. As a first approximation, it can still be regarded as gaussian, with the following parameters (approximate Rjanitsyn relations [13]):

$$\bar{x}_r = \bar{x} + 3.5\,(1 - 1/\sqrt[4]{r}\,)$$

$$\sigma_r = \sigma/\sqrt[4]{r}$$

b) If $F(x)$ is an extreme type I distribution, $F_r(x)$ is still extreme of type I, with mean and standard deviation:

$$\bar{x}_r = \bar{x} + (\ln r)/a$$

$$\sigma_r = \sigma$$

c) If $F(x)$ is an extreme type II distribution, $F_r(x)$ is also extreme of type II with parameters:

$$\bar{x}_r = \bar{x}\, r^{1/\beta}$$

$$c_r = c\ .$$

5.4 MULTIDIMENSIONAL VARIABILITY

5.4.1 Joint Distribution of n Random Variables

The expression:

$$F_{\underline{X}}(x_1, x_2, \ldots, x_n) = P \begin{Bmatrix} X_1 \leqslant x_1 \\ X_2 \leqslant x_2 \\ \ldots\ldots \\ X_n \leqslant x_n \end{Bmatrix} \qquad (5.48)$$

is defined as the joint distribution function of the random variables $X_1, X_2, \ldots, X_n$.

The joint probability density of the n variables is obtained from (5.48) through derivation:

$$f_{\underline{X}}(x_1, x_2, \ldots, x_n) = \frac{\partial^n F_{\underline{X}}(x_1, x_2, \ldots, x_n)}{\partial x_1 \, \partial x_2 \ldots \partial x_n} \qquad (5.49)$$

Moreover:

$$f_{\underline{X}}(x_1, x_2, \ldots, x_n)dx_1 \, dx_2 \ldots dx_n = P \left\{ \begin{array}{l} x_1 < X_1 \leqslant x_1 + dx_1 \\ x_2 < X_2 \leqslant x_2 + dx_2 \\ \ldots \ldots \ldots \ldots \\ x_n < X_n \leqslant x_n + dx_n \end{array} \right\} \quad (5.50)$$

The probability that the determination $\underline{X}$ belongs to the volume V defined in the n-dimensional hyperspace is:

$$P\{\underline{X} \in V\} = \int_V f_{\underline{X}}(x_1, x_2, \ldots, x_n)dx_1 \, dx_2 \ldots dx_n$$

and the normalisation condition becomes:

$$\int_{-\infty}^{+\infty} \ldots \int_{-\infty}^{+\infty} f_{\underline{X}}(x_1, x_2, \ldots, x_n)dx_1 \, dx_2 \ldots dx_n = 1 \qquad (5.51)$$

$$\underbrace{\qquad\qquad\qquad\qquad}_{n \text{ times}}$$

If (5.49) is known, then the joint probability density of k from among the n random variables can be easily determined.

Assuming $p = n - k$, it follows that:

$$f(x_1, x_2, \ldots, x_k) = \int_{-\infty}^{+\infty} \ldots \int_{-\infty}^{+\infty} f(x_1, x_2, \ldots, x_n)dx_{k+1} \ldots dx_n \qquad (5.52)$$

$$\underbrace{\qquad\qquad\qquad}_{p \text{ times}}$$

If the n random variables are stochastically independent, the multiplication theorem of probability calculation permits us to write:

$$f(x_1, x_2, \ldots, x_n) = \prod_{i=1}^{n} f(x_i) \qquad (5.53)$$

Thus, if $X_1, X_2, \ldots, X_n$ are independent and normally distributed with means $\bar{x}_i$ and standard deviations σ_i, we get:

$$f(x_1, x_2, \ldots, x_n) = \prod_{i=1}^{n} \frac{1}{\sigma_k \sqrt{2\pi}} \exp\left[-\frac{1}{2} \frac{(x_i - \bar{x}_i)^2}{\sigma_i^2} \right] =$$

$$\frac{1}{\sigma_1 \sigma_2 \ldots \sigma_n \sqrt{(2\pi)^n}} \exp\left[-\frac{1}{2} \sum_{i=1}^{n} \frac{(x_i - \bar{x}_i)^2}{\sigma_i^2} \right] \qquad (5.54)$$

5.4.1.1 Parameters of the joint n-dimensional distribution

The major parameters of a multidimensional distribution are:
—the means of the marginal variables X_i

$$\bar{x}_i = \int_{-\infty}^{+\infty} x_i \, f(x_i) \, dx_i \quad , \qquad (5.55)$$

—the variances of the marginal variables X_i

$$\sigma_{X_i}^2 = \sigma_{X_i X_i} = \int_{-\infty}^{+\infty} (x_i - \bar{x}_i)^2 \, f(x_i) \, dx_i \quad , \qquad (5.56)$$

—the covariances

$$\sigma_{X_i X_j} = \int_{-\infty}^{+\infty} \int_{-\infty}^{+\infty} (x_i - \bar{x}_i)(x_j - \bar{x}_j) \, f(x_i, x_j) \, dx_i \, dx_j \qquad (5.57)$$

The linear correlation coefficients $\rho_{X_i X_j}$ between the X_i and X_j variables are given by:

$$\rho_{X_i X_j} = \frac{\sigma_{X_i X_j}}{\sigma_{X_i} \sigma_{X_j}}$$

The covariance matrix:

$$[K] = \begin{bmatrix} \sigma_{X_1}^2 & \sigma_{X_1 X_2} & \cdots & \sigma_{X_1 X_n} \\ \sigma_{X_2 X_1} & \sigma_{X_2}^2 & \cdots & \sigma_{X_2 X_n} \\ \cdots\cdots & & & \\ \sigma_{X_n X_1} & \sigma_{X_n X_2} & \cdots & \sigma_{X_n}^2 \end{bmatrix}$$

is called the correlation matrix and illustrates the statistical dependence between the variables under consideration. If the variables X_i, X_j are independent, then covariance $\sigma_{X_i X_j}$ ($i \neq j$) is zero. In this case, in fact, on the basis of (5.53), we get:

$$f(x_i, x_j) = f(x_i)\, f(x_j)$$

and from (5.57):

$$\sigma_{X_i X_j} = \left[\int_{-\infty}^{+\infty} x_i\, f(x_i)\, dx_i - \bar{x}_i \right] \left[\int_{-\infty}^{+\infty} x_j\, f(x_j)\, dx_j - \bar{x}_j \right] = 0$$

By means of exp. (5.57) we can also deduce that the correlation matrix is symmetrical. Its diagonal elements $\sigma_{X_i X_j} = \sigma_{X_i}^2$ are the variances of the corresponding variables. For example, if the correlation matrix of the normal random variables $X_1, X_2, \ldots X_n$ is known, it is possible to write their joint probability density in the form:

$$f_{\underline{X}}(x_1, x_2, \ldots, x_n) =$$

$$\frac{1}{\sqrt{(2\pi)^n |K|}} \exp\left[-\frac{1}{2} \sum_{i=1}^{n} \sum_{j=1}^{n} \{K^{-1}\}_{ij} (x_i - \bar{x}_i)(x_j - \bar{x}_j) \right] \quad (5.58)$$

where:

$|K|$ denotes the determinant of the correlation matrix
$\{K^{-1}\}_{ij}$ denotes the generic element of the matrix K^{-1}, inverse of K.

In absence of correlation between the variables, the matrices K and K^{-1} become diagonal. In this case, one gets:

$$|K| = \sigma_1^2\, \sigma_2^2 \ldots \sigma_n^2$$

$$\{K^{-1}\}_{ij} = \begin{cases} 1/\sigma_i^2 & \text{for } i = j \\ 0 & \text{for } i \neq j \end{cases}$$

and (5.58) becomes (5.54).

5.5 FUNCTIONS OF THE RANDOM VARIABLES

Let the random variables $X_1, X_2, \ldots, X_n$ be given, and also the functional relation:

$$y = g(x_1, x_2, \ldots, x_n) \quad (5.59)$$

We want to determine the cumulative distribution function $F_Y(y)$, the joint distribution function $f_{\underline{X}}(x_1, x_2, \ldots, x_n)$ being known.

Denoting by D_{ny} the domain in the n-dimensional space for which

$$g(x_1, x_2, \ldots, x_n) \leqslant y$$

is valid, the cumulative distribution function of Y is given by:

$$F_Y(y) = P\{Y \leqslant y\} = P\{\underline{X} \in D_{ny}\} = \int_{D_{ny}} f_{\underline{X}}(x_1, x_2, \ldots, x_n) dx_1 \, dx_2 \ldots dx_n \tag{5.60}$$

Thus in the case of only two variables, (5.59) becomes:

$$y = g(x_1, x_2) \ .$$

Denoting by D_{2y} the domain, even not simply connected (Fig.5.2), such that:

$$g(x_1, x_2) \leqslant y \ ,$$

we get:

$$F_Y(y) = \iint_{D_{2y}} f_{X_1 X_2}(x_1, x_2) \, dx_1 \, dx_2 \tag{5.61}$$

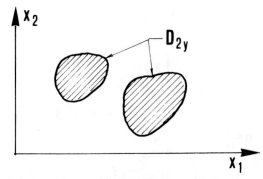

Fig. 5.2

For example, let:

$$y = x_1 + x_2 \qquad (-\infty \leqslant x_1 \leqslant \infty; \ -\infty \leqslant x_2 \leqslant \infty)$$

The region D_{2y} is constituted by the semiplane to the left of the straight line $x_1 + x_2 = y$.

Integrating by horizontal strips we get:

$$F_Y(y) = \int_{-\infty}^{+\infty} dx_2 \left[\int_{-\infty}^{y-x_2} f(x_1, x_2)dx_1 \right]$$

$$f_Y(y) = \frac{d\,F_Y(y)}{dy} = \int_{-\infty}^{+\infty} f(y-x_2, x_2)dx_2$$

Integrating by vertical strips we get:

$$F_Y(y) = \int_{-\infty}^{+\infty} dx_1 \left[\int_{-\infty}^{y-x_1} f(x_1, x_2)dx_2 \right]$$

$$f_Y(y) = \int_{-\infty}^{+\infty} f(x_1, y-x_1)dx_1$$

If X_1 and X_2 are independent $f(x_1, x_2) = f(x_1)\,f(x_2)$, the convolution relation obtained is:

$$f_Y(y) = \int_{-\infty}^{+\infty} f_{X_1}(y-x_2)f_{X_2}(x_2)dx_2$$

$$= \int_{-\infty}^{+\infty} f_{X_1}(x_1)\,f_{X_2}(y-x_1)dx_1 \tag{5.62}$$

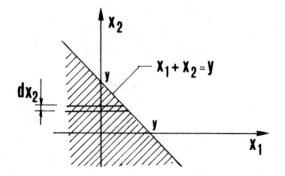

Fig. 5.3

In particular, if the variables X_1 and X_2 take only positive values, (5.62) becomes:

$$f_Y(y) = \int_0^y f_{X_1}(y-x_2) f_{X_2}(x_2) dx_2$$

$$= \int_0^y f_{X_1}(x_1) f_{X_2}(y-x_1) dx_1 \tag{5.63}$$

In the case where Y is a function of only one random variable:

$$y = g(x) \tag{5.64}$$

the integration domain D_{1y} reduces to one or more segments. Hence, for any functional relation, we get:

$$F_Y(y) = \int_{D_{1y}} f_X(x)\, dx$$

If (5.64) is monotone, let $x = h(y)$ be the inverse function. Then:

$$F_Y(y) = P\{Y \leqslant y\}$$

can be written in the forms:

$$F_Y(y) = \begin{cases} P\{X \leqslant h(y)\} = F_X[h(y)] & \text{for } g(\cdot) \text{ increasing} \\ P\{X \geqslant h(y)\} = 1 - F_X[h(y)] & \text{for } g(\cdot) \text{ decreasing} \end{cases}$$

from which, by derivation, one gets:

$$f_Y(y) = f_X[h(y)]\, |h'(y)| . \tag{5.65}$$

5.5.1 Joint Probability Density of k Functions of n Random Variables

Given k equations

$$y_1 = g_1(x_1, x_2, \ldots, x_n)$$
$$y_2 = g_2(x_1, x_2, \ldots, x_n) ,$$
$$\cdots \cdots$$
$$y_k = g_k(x_1, x_2, \ldots, x_n) \tag{5.66}$$

which relate the k dependent variables Y_i to the n basic variables X_j.

The system (5.66) admits of the solution:

$$x_1 = h_1(y_1, y_2, \ldots, y_k, x_{k+1}, \ldots, x_n)$$
$$x_2 = h_2(y_1, y_2, \ldots, y_k, x_{k+1}, \ldots, x_n) \qquad (5.67)$$
$$\cdots \cdots$$
$$x_k = h_k(y_1, y_2, \ldots, y_k, x_{k+1}, \ldots, x_n)$$

We want to determine the joint distribution function $f_Y(y_1, y_2, \ldots, y_k)$ of the variables Y_i, knowing the joint distribution function $f_X(x_1, x_2, \ldots, x_n)$ of the basic variables X_j.

Two cases commonly referred to in practical applications can be distinguished:

1st case: k = n

Denoting by:

$$J = \frac{\partial(h_1, h_2, \ldots, h_k)}{\partial(y_1, y_2, \ldots, y_k)} = \begin{vmatrix} \dfrac{\partial h_1}{\partial y_1} & \dfrac{\partial h_2}{\partial y_1} & \cdots & \dfrac{\partial h_k}{\partial y_1} \\ \dfrac{\partial h_1}{\partial y_2} & \dfrac{\partial h_2}{\partial y_2} & \cdots & \dfrac{\partial h_k}{\partial y_2} \\ \cdots \cdots \\ \dfrac{\partial h_1}{\partial y_k} & \dfrac{\partial h_2}{\partial y_k} & \cdots & \dfrac{\partial h_k}{\partial y_k} \end{vmatrix}$$

the jacobian determinant of the system of functions (5.67), the extension of (5.65) to the multidimensional case yields:

$$f_Y(y_1, y_2, \ldots, y_k) = f_X(h_1, h_2, \ldots, h_k) \left| \frac{\partial(h_1, h_2, \ldots, h_k)}{\partial(y_1, y_2, \ldots, y_k)} \right| \qquad (5.68)$$

2nd case: k < n (n − k = p)

Operating in a way analogous to the first case, bearing in mind (5.67) and (5.52), we get:

$$f_Y(y_1, y_2, \ldots, y_k) = \int_{-\infty}^{+\infty} \ldots \int_{-\infty}^{+\infty} f_X(h_1, \ldots, h_k, x_{k+1}, \ldots, x_n) \cdot$$
$$\underbrace{\qquad\qquad}_{p \text{ times}} \qquad (5.69)$$

$$\cdot \quad \frac{\partial(h_1, h_2, \ldots, h_k)}{\partial(y_1, y_2, \ldots, y_k)} \quad dx_{k+1} \cdots dx_n$$

5.6 LINEAR TRANSFORMATIONS OF NORMAL VARIABLES

Normal distributions have the important property of remaining normal even under linear transformations of the individual random variables.

In the one-dimensional case, let X be a normal variable, with probability density:

$$f_X(x) = \frac{1}{\sigma_X \sqrt{2\pi}} \exp\left[-\frac{(x - \bar{x})^2}{2\sigma_X^2}\right] \qquad (5.70)$$

We want to determine the probability density of the random variable Y connected to X by the linear relationship:

$$y = ax + c \ .$$

From the relationship:

$$f_Y(y) = f_X[h(y)] \ |h'(y)|$$

where: $x = h(y) = (y - c)/a$, we get:

$$f_Y(y) = \frac{1}{a\sigma_X\sqrt{2\pi}} \exp\left[-\frac{(y - c - a\bar{x})^2}{2 a^2 \sigma_X^2}\right]$$

Writing:

$$\bar{y} = a \bar{x} + c \ ,$$

$$\sigma_Y = a \sigma_X \ ,$$

from (5.70) we deduce that the variable Y is normally distributed with parameters $\bar{y}$ and σ_Y.

Now let:

$$y_j = \sum_{k=1}^{n} a_{jk} x_k + c_j \qquad (j = 1, 2, \ldots, n)$$

be n functional relations and X_k (k = 1, 2, $\ldots$, n) n normal variables with joint probability density:

$$f_{\underline{X}}(x_1, x_2, \ldots, x_n) =$$

$$\frac{1}{\sqrt{(2\pi)^n |K_X|}} \exp\left[-\frac{1}{2} \sum_{j=1}^{n} \sum_{k=1}^{n} \{K_X^{-1}\}_{jk} (x_j - \bar{x}_j)(x_k - \bar{x}_k)\right]$$

It follows that:

$$f_{\underline{Y}}(y_1, y_2, \ldots, y_n) =$$

$$\frac{1}{\sqrt{(2\pi)^n |K_Y|}} \exp\left[-\frac{1}{2} \sum_{j=1}^{n} \sum_{k=1}^{n} \{K_Y^{-1}\}_{jk} (y_j - \bar{y}_j)(y_k - \bar{y}_k)\right]$$

with:

$$\bar{y}_j = \sum_{k=1}^{n} a_{jk} \bar{x}_k + c_j \qquad (5.71)$$

$$\sigma_{Y_i Y_j} = E[(y_j - \bar{y}_j)(y_k - \bar{y}_k)] = E\left[\sum_{a=1}^{n} a_{ja}(x_a - \bar{x}_a) \sum_{\beta=1}^{n} a_{k\beta}(x_\beta - \bar{x}_\beta)\right]$$

$$= \sum_{a=1}^{n} \sum_{\beta=1}^{n} a_{ja} a_{k\beta} \sigma_{x_a x_\beta} \qquad (5.72)$$

The last two relations above remain valid not only for linear transformations of normal variables but also for those of variables of any distribution; however, they do not hold for nonlinear transformations of any variables, not even for normal variables.

In that case, the functional relationships:

$$y_j = g_j(x_1, x_2, \ldots, x_n)$$

can be replaced, as a first approximation, by their Taylor power series expansions in the neighbourhood of the mean values, neglecting nonlinear terms.

5.7 APPROXIMATE DERIVATION OF THE DISTRIBUTIONS OF FUNCTIONS OF RANDOM VARIABLES. BRIEF NOTES ON THE MONTE CARLO METHOD

Let the random variables $X_1, X_2, \ldots, X_n$ and the functional relation $y = \varphi(x_1, x_2, \ldots, x_n)$ be given.

It is possible to determine the approximate distribution of Y by means of simulation techniques of the Monte Carlo type. Their use is only subject to the one condition that the dependent variable Y be related in a unique manner to the basic variables X_i, while no restriction is imposed on the number of the variables X_i, on their probability densities, on the existence or nonexistence of any correlation between the variables X_i, on the type of dependence between Y and the variables X_i. Among other things, it is not necessary that the function $\varphi(\cdot)$ be expressed explicitly; it is sufficient to know the numerical procedure that permits us to obtain Y from the X_i variables.

The method consists of the simulation of the random character of each basic variable X_i by means of the artificial generation of m random numbers whose frequency distribution converges to $f_{X_i}(x_i)$ as the number m of the experiments increases.

Simulating all the variables involved, the matrix of the data can be produced, having n rows and m columns. Because of the postulated unique dependence, a unique value of the dependent variable Y corresponds to the elements of any column of this matrix.

Making use of all the columns, we get a sample composed of m argumental values of Y, which permits the plotting of a frequency histogram and hence the desired approximate probability distribution.

To obtain accurate information concerning the tails of the distribution of Y, as required in the context of level 3 operational procedures, it is necessary to perform a considerable number of experiments (10^6 to 10^9). This number is significantly reduced if research is limited to finding the estimated central values of the distribution.

5.8 ORTHOGONAL TRANSFORMATIONS OF RANDOM VARIABLES

In various structural safety problems it is useful to have the possibility of operating with variables that are uncorrelated, i.e.

to have zero covariances. Given the random variables $X_1, X_2, \ldots,$ X_n with means $\bar{x}_1, \bar{x}_2, \ldots, \bar{x}_n$ and correlation matrix:

$$[K_X] = \begin{vmatrix} \sigma_{11} & \sigma_{12} & \cdots & \sigma_{1n} \\ \sigma_{21} & \sigma_{22} & \cdots & \sigma_{2n} \\ \cdots\cdots \\ \sigma_{n1} & \sigma_{n2} & \cdots & \sigma_{nn} \end{vmatrix} \qquad (5.73)$$

one must then determine n variables $Y_1, Y_2, \ldots, Y_n$ such that their correlation matrix $[K_Y]$ will be diagonal, that is of the type:

$$[K_Y] = \begin{vmatrix} \sigma'_{11} & 0 & \cdots & 0 \\ 0 & \sigma'_{22} & \cdots & 0 \\ \cdots\cdots \\ 0 & 0 & \cdots & \sigma'_{nn} \end{vmatrix} \qquad (5.74)$$

in which $\sigma'_{kk} = (\sigma'_k)^2$ denotes the variances of the variables Y_k.

These variances σ'_{kk} are the eigenvalues of matrix $[K_X]$; the variables Y_k are expressed by the linear functional relationships:

$$\underline{y} = [U^T] \underline{x} \qquad (5.75)$$

where $[U^T]$ is the transposed matrix of the eigenvectors relative to the eigenvalues of $[K_X]$. $[K_X]$ being symmetrical, the transposed matrix and the inverse matrix of the eigenvectors are identical.

Therefore the variances σ'_{kk} $(k = 1, 2, \ldots, n)$ of the uncorrelated variables Y_k can be deduced by solving the characteristic equation:

$$\begin{vmatrix} (\sigma_{11}-\sigma') & \sigma_{12} & \cdots & \sigma_{1n} \\ \sigma_{21} & (\sigma_{22}-\sigma') & \cdots & \sigma_{2n} \\ \cdots\cdots \\ \sigma_{n1} & \sigma_{n2} & \cdots & (\sigma_{nn}-\sigma') \end{vmatrix} = 0 \quad (5.76)$$

that is, the algebraic equation:

$$(\sigma')^n - (\sigma_{11} + \sigma_{22} + \ldots + \sigma_{nn})(\sigma')^{n-1} + \ldots + (-1)^n \det|\sigma_{ij}| = 0$$

The eigenvectors a_{ik} are determined by solving the n homogeneous systems:

$$\left|\begin{array}{l} (\sigma_{11}-\sigma'_{kk})a_{1k} + \sigma_{12}a_{2k} + \ldots + \sigma_{1n}a_{nk} = 0 \\ \sigma_{21}a_{1k} + (\sigma_{22}-\sigma'_{kk})a_{2k} + \ldots + \sigma_{2n}a_{nk} = 0 \\ \ldots\ldots \\ \sigma_{n1}a_{1k} + \sigma_{n2}a_{2k} + \ldots + (\sigma_{nn}-\sigma'_{kk})a_{nk} = 0 \end{array}\right| \qquad (5.77)$$

$$(k = 1, 2, \ldots, n)$$

keeping in mind, for every fixed k, the normalisation condition:

$$\sum_{i=1}^{n} a_{ik}^2 = 1 \ .$$

The transposed matrix $[U^T]$ of the eigenvectors will be:

$$[U^T] = \left|\begin{array}{cccc} a_{11} & a_{21} & \cdots & a_{n1} \\ a_{12} & a_{22} & \cdots & a_{n2} \\ \ldots\ldots \\ a_{1n} & a_{2n} & \cdots & a_{nn} \end{array}\right| \qquad (5.78)$$

and the desired functional relationships are:

$$y_1 = a_{11}x_1 + a_{21}x_2 + \ldots + a_{n1}x_n$$
$$y_2 = a_{12}x_1 + a_{22}x_2 + \ldots + a_{n2}x_n$$
$$\ldots\ldots$$
$$y_n = a_{1n}x_1 + a_{2n}x_2 + \ldots + a_{nn}x_n$$

$$(5.79)$$

Hence, from the theoretical viewpoint, the problem of eliminating the correlation between the variables is reduced to finding the n roots of an algebraic equation and solving n homogeneous systems of linear equations. Actually, when n is greater than 3, in practical computation it is not easy to apply the direct method examined above. In these cases it is possible to resort to specific operational procedures, such as for instance the Jacobi method[70].

References

[1] Forsell, C. "Ekonomioch Byggnadsvasen", Sunt Fomoft, 1924

[2] Mayer, H. "Die Sicherheit der Bauwerke", Springer Verlag, Berlin, 1926

[3] Freudenthal, A.M. "Safety and Probability of Structural Failure", Transactions ASCE, vol. 121, 1956

[4] Paez, A. – Torroja, E. "La Determination del Coeficiente de Seguridad en las Distintas Obras", Instituto Tecnico de la Construccion y del Cemento, Madrid, 1959

[5] Johnson, A.I. "Strength, Safety and Economical Dimension of Structures", Bulletin n° 12, Royal Institute of Technology, Stockholm, 1953

[6] Streletzki, N.S. "Osnovy Statisticheskogo Uchota Koefficienta Zapesa Prochnosti Sooruzhenyi", Stroizdat, Moskva, 1947

[7] Veneziano, D. "Basic Principles and Methods of Structural Safety", MIT, Research Report R 76 – 2, 1976

[8] Von Neumann, J. – Morgenstern, O. "The Theory of Games and Economic Behaviour", Princeton University Press, 3rd ed., 1953

[9] Rosenblueth, E. "Code Specification on Safety and Service-ability", Planning and Design of Tall Buildings, Technical Committee 10, ASCE – IABSE, 1972

[10] CEB Bulletin n° 116–E, vol.I, Paris, 1976

[11] Prager,W. "An Introduction to Plasticity", Addison–Wesley, 1959

[12] Bolotin,V.V. "Application of Methods of the Theory of Probability in the Theory of Plates and Shells", Proceedings of the 4th All-Union Conference on Shells and Plates, Erevan (URSS), 1962

[13] Rjanitsyn, A.R. "Calcul à la rupture et plasticité des con-structions", Eyrolles – Paris, 1959

[14] Freudenthal, A.M.–Garrelts, J.M.–Shinozuka, M. "The Ana-lysis of Structural Safety", Journal of the Structural Division, Proceedings of ASCE, February 1966

[15] Gavarini, C. – Veneziano, D. "Sulla teoria probabilistica degli stati limite delle strutture", Giornale del Genio Civile, n.ri 11–12, 1970

[16] Borges, J. Ferry–Castanheta, M. "Generalized Theory of Structural Safety", C 1–1, LNEC, Lisbon, 1973

[17] Tichy, M.–Vorlicek, M. "Statistical Theory of Concrete Structures", Academia, Prague, 1972

[18] Benjamin, J.R.–Cornell, C.A. "Probability, Statistics and Decision for Civil Engineers", McGraw-Hill Book Co., 1970

[19] Tribus, M. "Rational Descriptions, Decisions and Design", Pergamon Press, 1969

[20] Weibull, W. "Fatigue Testing and Analysis of Results", Pergamon Press, 1961

[21] CEB Bulletin d'information n° 112, Paris, 1976

[22] Borges, J. Ferry—Castanheta, M. "Structural Safety", LNEC, Lisbon, 1971

[23] Cornell, C.A. "A Proposal for a Probability-based Code Suitable for Immediate Implementation", Memorandum to ASCE and ACI Committees on Structural Safety, MIT, Cambridge, 1967

[24] Galambos, T.V.—Ravindra, M.K. "Probability-based Load Factor Design Criteria for Steel Beams", Proceedings of the ASCE Specialty Conference on the Safety and Reliability of Metal Structures, Pittsburg, 1972

[25] Parimi, S.R.—Lind N.C. "Limit States Basis for Cold-formed Steel Design", Journal of the Structural Division, March 1976

[26] Esteva, L.—Rosenblueth, E. "Use of Reliability Theory in Building Codes", Conference on Applications of Statistics and Probability to Soil and Structural Engineering, University of Hong Kong, 1971

[27] Nordic Committee for Building Regulations—Sub-Committee on Structural Safety "Proposal for Safety Codes for Load-carrying Structures", 1974

[28] Allen, D.E. (translator) "A Statistical Method of Design of Building Structures", Translation of five Russian articles, Tech. Trans. 1368, N.R.C. of Canada, 1969

[29] Basler, E. "Untersuchungen über den Sichereitsbegriff von Bauwerken", Schweitzer Archiv N° 4, vol. 27, 1961

[30] Cornell, C.A. "Structural Safety Specification based on Second-Moment Reliability Analysis", IABSE Symposium, London, 1969

[31] Rosenblueth, E.—Esteva, L. "Reliability Basis for Some Mexican Codes", ACI Publication SP—31, 1971

[32] Ditlevsen, O. "Structural Reliability and the Invariance Problem", Report n° 22, Solid Mechanics Division, University of Waterloo, Canada, 1973

[33] Hasofer, A.M.–Lind, N.C. "Exact and Invariant Second-Moment Code Format", Journal of the Engineering Mechanics Division, February 1974

[34] Veneziano, D. "Second Moment Reliability", Research Report R 74–33, MIT, Cambridge, 1974

[35] Veneziano, D. "Statistical Inference in Second-Moment Reliability", Research Report R 74–33, MIT, Cambridge, 1974

[36] Ditlevsen, O. "A Collection of Notes Concerning Structural Reliability", Dialog, 2–76, Denmarks Ingeniørakademi, Bygningsafdelingen, Lyngby, 1976

[37] Veneziano, D. "Safe Regions and Second-Moment Reliability Vectors", Research Report R 74–33, MIT, Cambridge, 1974

[38] Ditlevsen, O.–Skov, K. "Uncertainty Theoretical Definition of Structural Safety suggested by the NKB–Safety Committee", Dialog, 2–76, Denmarks Ingeniørakademi, Bygningsafdelingen, Lyngby, 1976

[39] Lind, C. "A Formulation of Probabilistic Design", paper n° 128, Solid Mechanics Division, University of Waterloo, 1974

[40] Fiessler,B.–Rackwitz,R. "Note on Discrete Safety Checking when Using Non-normal Stochastics Models for Basic Variables", Load Project Working Session, Massachusetts Institute of Technology, Cambridge, 1976

[41] Paloheimo,E. "Eine Bemessungsmethode die sich auf variierende Fraktile Gründet", Sicherheit von Betonbauten, D.B.V., 1973

[42] Paloheimo,E.–Hannus,H. "Structural Design Based on Weighted Fractiles", Journal of the Structural Division, 1974

[43] Ang,A. H–S. "A Comprehensive Basis for Reliability Analysis and Design", Reliability Approach in Structural Engineering, Maruzen Co., Ltd, Tokyo, 1975

[44] Rackwitz,R. "Principles and Methods for a Practical Probabilistic Approach to Structural Design", Technical University of Munich, 1975

[45] Turkstra,C.J. "Theory of Structural Design Decision", Solid Mechanics Division Study n° 2, University of Waterloo, Canada, 1970

[46] Tichy,M. "A Probabilistic Model for Structural Actions", Acta Techniča ČSAV n° 5, Praha, 1974

[47] Bredsdorff,W.–Kukulski,W.–Skov,K. "Recommendation on General Principles of Structural Design", Second Preliminary Draft of a Report to Economic Commission for Europe –Committee on Housing, Building and Planning, SBI, Denmark, 1976

[48] CEB Bulletin d'Information n° 72 "Recommandations Internationales CEB–FIP pur le calcul et l'exécution des ouvrages en beton", Principes et Recommandations, Paris, 1970

[49] International Standards Organisation "General Principles for the Verification of the Safety of Structures" ISO 2394, 1973

[50] CEB Bulletin n° 111 "Système International de Reglementation Technique Unifiée des Structures", Paris, 1975

[51] CECM–ECCS "Recommendations for Steel Structures–European Convention for Constructional Steelwork", Draft, 1975

[52] American Iron and Steel Institute "Proposed Criteria for Load and Resistance Factor Design of Steel Building Structures", Project 163, Preliminary draft, 1975

[53] Lind,N.C. "Application to Design of Level 1 Codes", CEB Bulletin n° 112, 1976

[54] Baker,M.J. "Evaluation of Partial Safety Factors for Level 1 Codes–Example of Application of Methods to Reinforced Concrete Beams", CEB Bulletin n° 112, 1976

[55] Leporati,E. "I valori caratteristici nella sicurezza strutturale e nel controllo della qualità", Costruzioni Metalliche, n° 5, 1974

[56] Borges,J. Ferry–Castanheta,M. "Generalized Theory of Structural Safety", CEB International Course on Structural Concrete C 1–1, L.N.E.C., Lisbon, 1973

[57] Borges,J. Ferry "The Checking of Safety of Non Linear Structures", Symposium on Non Linear Techniques and Behaviour in Structural Analysis, Crowthorne, 1974

[58] Levi,F.–Leporati,E. "Sulla scelta delle modalità operative nelle verifiche strutturali non lineari", Accademia Nazionale dei Lincei, Rendiconti, Serie VIII; Nota 1, marzo 1975; Nota II, aprile 1975

[59] Mathieu,H. Manuel "Securité des Structures"–concepts generaux, charges et actions, CEB Bulletin d'Information n° 106–107, 1975

[60] Bulletin d'Information CEB n° 117–F "Code Modele pour les Structures en Beton", volume II–Paris, Decembre 1976

[61] Skov,K. "A Comparison between the Principle of Level 2 and Level 1 and a Method for the Determination of Level 1 Safety Elements", J.C.S.S., Paris, November 1976

[62] Giuffrè,N.–Pinto,P.E. "Discretisation from a Level 2 Method", CEB Bulletin d'Information n° 112, 1976

[63] Pottharst,R. "Derivation of Level 1 Partial Safety Factors", Institut für Massivbau, Technische Hochschule Darmstadt, 1976

[64] Cramer,H. "Mathematical Methods of Statistics", Princeton University Press, 1971

[65] Gumbel,E.J. "Statistics of Extremes", Columbia University Press, New York, 1967

[66] VonMises,R. "Mathematical Theory of Probability and Statistics", Academic Press, New York, 1962

[67] Kolmogorov,A.N. "Grundbegriffe der Wahrscheinlicheitsrechnung", Springer, Berlin, 1933

[68] Fisz,M. "Probability Theory and Mathematical Statistics", Wiley, New York, 1963

[69] Godwin,H.J. "Generalisations of Tchebycheff's Inequality", Journal of American Statistical Association, 50, 923, 1955

[70] Fröberg,C.E. "Introduction to Numerical Analysis", Addison-Wesley Publishing Company, London, 1974

130

Notation

A	constant
a	integer
a	parameter
a_{ik}	eigenvector
B	event
B	Borel set
b	parameter
c	coefficient of variation
D	parameter
D	safety domain
D'	failure domain
E	expectation
$F(\cdot)$	cumulative distribution function
$F^*(x_i)$	cumulative distribution function given that action X_i occurs
F_i	i^{th} action
$f(\cdot)$	probability density
f_c	concrete strength
f_s	steel strength
$G(\cdot)$	conditional probability of failure
$g(\cdot)$	functional relation
i	integer
J	jacobian
j	integer
$h(\cdot)$	inverse function

K	covariance matrix
K^{-1}	inverse of matrix K
k	integer
k	coefficient
m_k	simple k-order moment
n	integer
p	integer
p_f	probability of failure
p_f^*	prefixed value of p_f
R	resistance
R_n	n-dimensional space
r	resistance function
r*	design value of variable R
r_i	independent repetition of variable action X_i
S	space of elementary events
S	load effect
s	load effect function
s*	design value of variable S
T	reference period
U	matrix of eigenvectors
U^T	transposed matrix of eigenvectors
U_i	i^{th} standardised variable
u_i	argumental value of U_i
V	variable
W	variable
w_i	predicted relative frequency of future demand
X_i	i^{th} basic variable
$\underline{X}$	basic variable vector
x_i	argumental value of X_i
x_i^*	design value of X_i
x_{ik}	characteristic value of X_i
$\bar{x}$	mean
$\check{x}$	median
$\tilde{x}$	mode
x_a	a-fractile
Y_i	i^{th} variable
y_i	argumental value of Y_i
Z	limit state variable
z	limit state function
a	integer
a_i	sensitivity factor, director cosine

β	integer
β	reliability index
Γ	gamma function
Γ	safety domain boundary
γ	coefficient of skewness
γ	safety factor
γ_o	central safety factor
γ_k	characteristic safety factor
$\gamma*$	design safety factor
γ_f	partial safety factor on loading
γ_m	partial safety factor on material strength
Δ	partial safety factor
$\underline{\delta}$	vector
δ_i	component of $\underline{\delta}$
η	standardised variable
ξ	standardised variable
ϑ	decision variable, design parameter
λ	constant
μ_k	central moment of order k
ρ	correlation coefficient
σ	standard deviation
σ_{ii}, σ^2	variance
σ_{ij}	covariance
σ'_{kk}	eigenvalue
$\emptyset$	empty set
$\emptyset(\cdot)$	standardised cumulative normal distribution
$\varphi(\cdot)$	standardised normal probability density
φ	functional relation
χ^2	cumulative chi-square distribution
ψ	partial factor on loading
ϵ	signifies belonging to
Π	signifies product